STUDY GUIDE

For

Berk

Development Through the Lifespan

Second Edition

STUDY GUIDE

For

Berk

Development Through the Lifespan

Second Edition

Laura E. Berk and Amy Petersen
Illinois State University

Allyn and Bacon
Boston London Toronto Sydney Tokyo Singapore

A Pearson Education Company
160 Gould Street
Needham Heights, MA 02494

Internet: www.abacon.com

ISBN 0-205-32143-7

Printed in the United States of America

10 9 8 7 6 5 4 3 2   05 04 03 02 01

CONTENTS

PREFACE

As you embark on the fascinating journey of studying human development, it is our hope that this Study Guide will help you master the material in your text, *Development Through the Lifespan.* Our intention in preparing the Study Guide is to provide you with active practice in learning the content of your textbook, thought-provoking questions that help you think critically about the material, and pleasurable exercises that assist you in mastering the basic vocabulary of the field. But to get the most out of the Study Guide, you must approach your reading assignments and classroom experiences with good study habits. In the following sections, we offer some basic study tips, which, if followed carefully, will improve your learning in any college course. Then we show how you can use this Study Guide to enhance your mastery of human development.

STUDY TIPS

Learning any new subject is a challenging and effortful task. Organizing your time, creating a climate for learning, previewing the material, reading actively, reviewing effectively, and using effective classroom learning techniques are essential parts of any successful study plan.

1. Organizing Your Time. Map out your study hours ahead of time to guarantee that you have enough time to study well. Purchase a study plan book covering the school year in which the days of the week are divided into time periods with sufficient space to write in a weekly schedule. Set aside at least two hours of study time for each class meeting. At the beginning of each week, write in what you intend to do during each study period. Then stick to your plan, checking off each item listed as you complete it. At the end of each week, evaluate how well your study schedule worked, and modify the next week's plan in light of your conclusions.

Following a schedule eases the work of studying because it does away with the energy-consuming, conflict-ridden tasks of making decisions like these: "Should I study now or later?" "If I study now, what should I do first and what next?" In addition, your schedule will probably show that you have more time than you thought for activities you want to do—especially if you maintain your study schedule and use your time wisely.

2. Creating a Climate for Learning. Choose a place to study that is relatively free of distractions, both visual and auditory. If your desk is cluttered with photos and mementos, clear it so that you will not be tempted to focus on those items instead of your course assignments. To mask distracting sounds, some students find it helpful to play soft music while studying. If you do so, try to choose music that is not attention-grabbing itself. Make sure your study area is well lighted and that study implements (pens, pencils, and paper) are nearby so that you do not have to interrupt your work to get them. Finally, if you study regularly in a special place set aside for that purpose, you will save time because entering that space will trigger a "mental set" for getting down to work quickly.

3. Previewing the Material. Begin mastering a chapter by previewing the material. Look ahead and identify the highlights of the assignment by reading the introduction, major headings, and summaries. Previewing permits you to grasp the structure of the information in the assignment. When you know the layout of what you are about to learn, you are better able to organize the information meaningfully and remember it. Previewing has been shown to reduce the amount of study time needed while increasing learning.

4. Reading Actively. Read carefully, searching for main points. Underline them in your textbook and write them down in your notebook. Keep in mind that studying is different from just reading. When you study, your mind is constantly active, and you are always searching for important ideas. If you approach your assignment passively, by reading without grappling with the material, you will retain very little.

5. Reviewing Effectively. After you finish the assignment, review what you have read. Once again, read the chapter introduction, major headings, and summaries. However, keep in mind that reviewing is not limited to rereading. It involves trying to recall what you have read. Jot down key ideas associated with each major heading. Make up questions about the material and answer them, checking your answers against the content of your text. As you test yourself, you will be anticipating questions that your instructor will ask on examinations. Since you will have answered them in advance, exams will be that much easier when you take them.

6. Applying Effective Classroom Learning Techniques. Effective classroom learning requires good listening and note-taking skills. In class, exercise the utmost concentration and discipline in holding your mind on the track of your instructor. As you listen, become actively engaged with the subject matter. Try to anticipate the next point, summarize what has been said in your own words, and imagine the test questions that might be asked about the material. Offer comments about content related to your own knowledge and experience, and ask your instructor about statements that are unclear. The more involved you are in class, the easier it will be to remember the information.

Just as you listen actively, take notes energetically. The format of your notes is critically important. Put the date of the lecture and its subject at the top of the page. Write your notes toward the right-hand side, leaving a generous left-hand margin. Listen for key ideas, and write them down in an organized fashion; do not try to capture every word. As soon after the lecture as possible, read your notes and fill in the gaps. Use the left-hand margin to highlight major points and add more detail.

7. Before, During, and After an Examination. If you follow the suggestions in the preceding sections, you will avoid the frantic, last-minute cramming that results in disorganized thinking, poor retention, and fear of the test-taking situation. Instead, since you have learned the material gradually and methodically, you will be in a good position to approach the examination confidently—as an opportunity to show your instructor the extent of your knowledge. Get a good night's sleep beforehand, and review your notes one more time the day of the test.

When you get the examination, look it over to get an idea of how quickly you will need to work. Then budget your time accordingly. If the test is in objective format (such as multiple choice and true-false), read each question carefully and entirely, making sure that you notice all qualifying words, such as *usually, always, most, never,* and *some.* At the same time, refrain from reading into questions information that is not there. If the test is in essay format, it requires that you organize your thoughts and express them clearly to do well. Write a brief outline of major points in the margin of your paper, thinking carefully about the arrangement of ideas. As your write, be sure you stick closely to the subject of the question. Include examples, important facts, and explanations to document your generalizations; avoid empty padding of the answer. Reserve 5 to 10 minutes of the period to check your paper before you turn it in.

After the exam has been graded, look over your errors carefully, checking incorrect and omitted answers against the textbook and lecture notes. View the test results as an opportunity to find out what your weaknesses are and to take steps to overcome them. By analyzing the reasons for your mistakes, you will greatly reduce the chances of repeating them on the next text.

USING THE STUDY GUIDE

This Study Guide is designed to help you create an effective study system for mastering human development. Each of its chapters is organized into the following parts:

CHAPTER SUMMARY

We begin with a brief summary of the chapter, mentioning major topics and general principles in text discussion. Each text chapter includes three additional summaries: (1) an informal summary at the beginning of each chapter, (2) a structured summary at the end of the chapter, and (3) a series of brief reviews, or interim summaries, placed at critical points in the text narrative. Thus, the summary in the Study Guide will be your fourth review of the information in the chapter and should be read before beginning the remaining activities in the guide.

LEARNING OBJECTIVES

The learning objectives describe what you should be able to do once you have mastered the material in each section of the chapter. Look over these objectives before you begin to read, as part of previewing. Then take notes on information relevant to each objective as you move through the chapter. When you finish reading, try to answer each objective in a few sentences or short paragraphs as you review. Once you have completed this exercise, you will have generated your own summary of chapter content. Because it will be written in your own words, it should serve as an especially useful chapter overview that can be referred to as you prepare for examinations.

STUDY QUESTIONS

The main body of each Study Guide chapter consists of study questions, organized according to major headings in the textbook, that will assist you in identifying main ideas and grasping essential concepts. You can use the study questions in several different ways. You may find it helpful to answer each question as you read the chapter, as part of your effort to read actively. Alternatively, try reading one or more sections and testing yourself by answering the relevant study questions. Finally, use the study question section as a device to review for examinations. If you work through it methodically, your retention of chapter material will be greatly improved.

ASK YOURSELF . . . QUESTIONS

In each chapter, we have included critical thinking questions that appear near brief summaries in your text book. Answering these questions will help you analyze important theoretical concepts and research findings. Many questions require that you apply what you have learned to problematic situations faced by parents, teachers, and children. In this way, they will help you think deeply about the material and inspire new insights. Each question is page-referenced to chapter material that will help you formulate a response.

TERM REVIEW PUZZLES

To help you master the central vocabulary of the field, we have provided crossword puzzles that test your knowledge of important terms and concepts. Answers can be found at the back of the Study Guide. If you cannot think of the term that matches a clue in the puzzles, your knowledge of information related to the term may be insecure. Reread the material in the text chapter related to each item that you miss. Also, try a more demanding approach to term mastery: After you have completed each puzzle, cover the definitions given in the clues and write your own explanation of each term.

SELF-TESTS

Once you have thoroughly studied each chapter, test your knowledge by taking the 25-item self-test. Then check your answers using the key at the back of the Study Guide. Each item is page-referenced to chapter content so that you can conveniently return to textbook sections related to questions that you missed. If you answered more than a few items incorrectly, spend extra time rereading the chapter, writing responses to learning objectives, and reviewing the study question section of this guide.

SUGGESTED READINGS

We hope that, as a result of studying human development, you will be motivated to extend your knowledge. In each Study Guide chapter, you will find from three to five suggested readings, which have been carefully chosen to build on chapter content and to be accessible to students who are new to the field of human development. Each suggested reading is accompanied by a brief description of its content.

Now that you understand how this Study Guide is organized, you are ready to begin using it to master *Development Through the Lifespan.* We wish you a rewarding and enjoyable course of study.

Acknowledgments

Our special thanks to Elizabeth Kenny for her invaluable assistance with the editing process.

Amy Petersen and Laura E. Berk

CHAPTER 1

HISTORY, THEORY, AND RESEARCH STRATEGIES

BRIEF CHAPTER SUMMARY

Human development is the study of all aspects of constancy and change throughout the lifespan. Theories lend structure and meaning to the scientific study of development. This chapter provides an overview of philosophical and theoretical approaches to the study of human development from medieval to modern times and reviews major research strategies used to study human behavior and development.

When compared and contrasted, historical philosophies and contemporary theories raise three basic questions about what people are like and how they develop: (1) Is development a continuous or discontinuous process? (2) Is there one course of development or many? (3) Are genetic or environmental factors more important in determining development? Although some theories take extreme positions on these issues, many modern ones include elements from both sides. The lifespan perspective recognizes that great complexity exists in human change and the factors that underlie it.

Research methods commonly used to study development include systematic observation, self-reports, clinical or case studies of single individuals, and ethnographies of the life circumstances of specific groups of people. Investigators of human development generally choose either a correlational or an experimental research design. To study how their subjects change over time, they apply special developmental research strategies—longitudinal, cross-sectional, and longitudinal-sequential designs. Each method and design has both strengths and limitations. Conducting research with human subjects also poses special ethical dilemmas.

Theory and research are the cornerstones of the field of human development. These components are helping us understand and alleviate many pressing problems faced by children and adults in today's world.

LEARNING OBJECTIVES

After reading this chapter, you should be able to:

1.1. Explain the importance of the terms *interdisciplinary* and *applied* as they help to define the field of human development. (5)

1.2. Explain the role of theories in understanding human development, and describe three basic issues on which major theories take a stand. (5-7)

1.3. Describe factors that sparked the emergence of the lifespan perspective, and explain the four assumptions that make up this point of view. (7-12)

1.4. Trace historical influences on modern theories of human development, from medieval times through the early twentieth century. (13-15)

1.5. Describe theoretical perspectives that influenced human development research in the mid-twentieth century, and cite the contributions and limitations of each. (16-21)

1.6. Describe recent theoretical perspectives on human development, noting the contributions of major theorists. (21-27)

1.7. Identify the stand that each contemporary theory takes on the three basic issues presented earlier in this chapter. (27-28)

1.8. Describe commonly used methods in research on human development, citing strengths and limitations of each. (28-31)

1.9. Contrast correlational and experimental research designs, and cite the strengths and limitations of each. (31-35)

1.10. Describe three research designs for studying development, and cite the strengths and limitations of each. (35-38)

1.11. Discuss special ethical concerns in lifespan research. (39-40)

STUDY QUESTIONS

Human Development as an Interdisciplinary, Scientific, and Applied Field

1. Describe two main factors that have stimulated the study of human development. (5)
A. ______________________________
B. ______________________________

2. Human development is an ______________________ field: it has grown from the combined efforts of people from many fields. (5)

Basic Issues

1. List three purposes of a good theory. (5)
A. ______________________ B. ______________________
C. ______________________

2. True or False: Theories differ from opinions and beliefs in that theories are subject to scientific verification. (6)

3. Match each theoretical approach with its description: (6-7)

_____	Assumes that all people follow the same sequence of development	1. Continuous theory
_____	Views development as adding on more of the same types of skills	2. Discontinuous theory
_____	Regards environment as the most important influence in development	3. Theory that stresses nature
_____	Assumes that children and adults live in distinct contexts	4. Theory that stresses nurture.
_____	Regards development as taking place in distinct stages	5. Theory that emphasizes one course of development
_____	Views heredity as the most important influence in development	6. Theory that emphasizes many courses of development

4. Theories that emphasize *stability* typically stress the importance of __________________ or ____________________________ in development. Other theorists believe that __________________ is likely if new experiences support it. (7)

The Lifespan Perspective: A Balanced Point of View

1. True or False: In taking a stand on the basic issues of human development, most modern theories recognize the importance of both sides of each issue. (7)

2. What trend, particularly striking in industrialized nations, has led researchers to move away from viewing adulthood and aging as periods of plateau and decline in development? (8)

____

3. List the eight periods of the lifespan covered in your text: (8)
A. ____
B. ____
C. ____
D. ____
E. ____
F. ____
G. ____
H. ____

4. The three interwoven domains of development to be examined within each period of the lifespan are: (9)
A. ____________________________ B. ____________________________
C. ____________________________.

5. The idea that development is *multidimensional* means that it is affected by an intricate blend of ______________________ and ____________________ forces. The idea that development is *multidirectional* means that it is a joint expression of ____________________ and ____________________ at all age periods. (9)

6. Development becomes less plastic, as ______________________ and ____________________ are reduced. (11)

7. The ability to adapt effectively in the face of adversity is ____________________. (10-11)

8. List three factors that offer individuals protection from stressful life events. (10-11)
A. ____
B. ____
C. ____

9. The idea that development is embedded in multiple contexts refers to the effects of what three types of influences? (11-12)
A. ____________________________ B. ____________________________
C. ____________________________

10. Nonnormative influences have become ____________ powerful and age-graded influences have become ____________ powerful in modern adult development. (12)

11. True or False: The lifespan perspective emphasizes multiple potential pathways and outcomes of development. (12)

Historical Foundations

1. Preformationism, a medieval view of development, regards children as ______________________________. (13)

2. Describe the influence of Puritanism on child-rearing values and practices during the Reformation. (13)

____

____

3. During the Enlightenment, the British philosopher John Locke regarded the child as a *tabula rasa,* which means ______________________________. According to this view, the child ____________________ contributes to his or her own development. (13-14)

4. What stand did Locke take on the three basic issues of child development? (14)
A. ______________________________ B. ______________________________
C. ______________________________

5. The French philosopher of the Enlightenment, Jean Jacques Rousseau, regarded the child as a *noble savage,* which means ______________________________
______________________________. (14)

6. Which two concepts found in modern theories does Rousseau's philosophy include? (14)
A. ______________________________ B. ______________________________

7. What stand did Rousseau take on the three basic issues of child development? (14)
A. ______________________________ B. ______________________________
C. ______________________________

8. What three areas of adult development were addressed by German philosopher John Nicolaus Tetens? (14)
A. ______________________________
B. ______________________________
C. ______________________________

9. German philosopher Friedrich August Carus identified four periods that span the life course:
A. ______________________________ B. ______________________________
C. ______________________________ D. ______________________________.
Like Tetens, Carus viewed aging not only as decline, but also as ______________________________
______________________________. (14)

10. Explain the two principles Darwin's theory of evolution emphasized. (14-15)
A. ______________________________

____

B. ______________________________

____

11. Charting age-related averages to represent the typical child's development is known as the ______________________________ to child study. Name the two major figures associated with this approach. (15)
A. ______________________________ B. ______________________________

12. What child-rearing advice is consistent with this view? (15)
____

13. ______________________________ constructed the first successful intelligence test, which became know as the ______________________________. (15)

14. This test provided ______________________________ and sparked interest in ______________________________. (15)

Mid-Twentieth-Century Theories

1. True or False: The psychoanalytic perspective emphasizes understanding the unique life history of each person. (16)

2. According to the psychoanalytic approach, people move through a series of stages in which they confront conflicts between ______________________________ and ______________________________. (16)

3. Match each of the following portions of the personality in Freud's theory with the appropriate description. (16-17)

_____	Source of basic biological needs and drives	1. Ego
_____	The conscious, rational part of personality	2. Superego
_____	Seat of conscience	3. Id

4. Match each of the following stages of psychosexual development with the appropriate description: (16)

_____	During this stage, sexual instincts die down	1. Genital
_____	Stage in which infant desires sucking activities	2. Anal
_____	Stage in which the Oedipal and Electra conflicts take place	3. Oral
_____	Stage marked by mature sexuality	4. Latency
_____	Stage in which toilet training becomes a major issue between parent and child	5. Phallic

5. Freud's theory highlighted the importance of ______________________________ and was the first to stress ______________________________.
However, his theory was criticized for what three reasons? (17)
A. ______________________________
B. ______________________________
C. ______________________________

6. While Freud's theory emphasized psycho______________ tasks of development, Erik Erikson's theory emphasized psycho______________ tasks. (17)

7. Erikson expanded upon Freud's vision by emphasizing that the ego is a ____________________ force in development, by adding ____________________ ____________________ to the model, and by pointing out that development must be understood in relation to ______________________________. (17)

8. Match each of Erikson's stages with its appropriate description: (18)

____ Successful resolution of this stage depends on the adult's success at caring for other people and productive work.	1. Industry vs. inferiority
____ The chief task of this stage is development of a coherent sense of who one is and one's place in society.	2. Autonomy vs. shame & doubt
____ Successful resolution of this stage depends on a warm, loving relationship with the caregiver.	3. Intimacy vs. isolation
____ In this stage, children experiment with adult roles through make-believe play.	4. Identity vs. identity diffusion
____ Successful resolution of this stage depends on parents granting the child reasonable opportunities for free choice.	5. Basic trust vs. mistrust
____ In this stage, successful resolution involves reflecting on life's accomplishments.	6. Generativity vs. stagnation
____ The development of close relationships with others helps ensure successful resolution of this stage.	7. Initiative vs. guilt
____ Children who develop the capacity for cooperation and productive work resolve this stage successfully.	8. Ego integrity vs. despair

9. Why is psychoanalytic theory no longer in the mainstream of human development research? (17-18))

____

____

10. True or False: Behaviorism focuses on the inner workings of the mind. (18)

11. Watson's study of little Albert, a 9-month-old baby who was taught to fear a white rat by associating it with a loud noise, supported Pavlov's concept of ______________________________. (19)

12. Skinner, who proposed ______________________________ theory, believed that behaviors could be increased by following them with ______________________ and decreased through ____________________. (19)

13. ____________________ devised the most influential social learning theory, which emphasizes the concept of ______________________________. (19)

14. Today, this theory emphasizes the importance of ____________________ and is called a ______________________________ rather than a social learning approach. (19)

15. Through observation and direct feedback, children develop ____________________ ____________________ and a sense of ____________________ which guide their responses in particular situations. (19)

16. What two concepts are combined in the use of *applied behavior analysis* to eliminate undesirable behaviors and increase socially acceptable ones? (19)
A. ______________________________ B. ______________________________

17. Behaviorism has been criticized for what two reasons? (19-20)
A. ______________________________
B. ______________________________

18. Briefly describe the view of children emphasized by Jean Piaget's cognitive-developmental theory. (20)
____
____

19. Match each of Piaget's stages with the appropriate description: (20-21)

_____	Thinking becomes abstract.	1. Preoperational
_____	Acting on the world with eyes, ears, and hands is the chief characteristic of this stage.	2. Concrete operational
_____	For the first time, children represent the world with symbols.	3. Formal operational
_____	Reasoning becomes logical and better organized.	4. Sensorimotor

20. Piaget used ______________________________ as his chief method for studying child and adolescent thought. (20)

21. Piaget's theory convinced the field that children are ______________________________ and sparked a wealth of research on ______________________________. (20-21)

22. List three recent challenges to Piaget's theory. (21)
A. ______________________________
B. ______________________________
C. ______________________________

Recent Theoretical Perspectives

1. According to the information-processing approach, the human mind is best viewed as ______________________________. (22)

2. Why do information-processing theorists use diagrams to represent mental operations? (22)
____
____

3. In what basic way are information processing and Piaget's theory alike? In what basic way are they different? (22)
A. ______________________________
B. ______________________________

4. Cite strengths and limitations of the information-processing approach. (23)
Strengths: ____
____
Limitations: ____
____

5. The modern foundations of ethology were laid by ______________________ and ______________________. (23)

6. What is meant by a *sensitive period?* How does it differ from the notion of a *critical period?* (23)
A. ____
____
B. ____
____

7. John Bowlby's ethological view of attachment suggests that attachment behaviors of babies, such as ______________________, are built-in social signals that help ensure that the baby will be ______________________. (23)

8. True or False: Ethology considers learning an important aspect of development. (24)

9. Cross-cultural and multi-cultural research helps us untangle the contributions of ______________ and ______________ factors in the timing, order of appearance, and diversity of behaviors. (24)

10. Describe how social interaction contributes to children's ability to guide their own actions in Vygotsky's sociocultural theory. (24)
____
____
____

11. List two ways Piaget and Vygotsky differ in their views of cognitive development. (24)
A. ____
____
B. ____
____

12. True or False: Because cultures select tasks for their members, individuals in every culture develop unique strengths not present in others. (25)

13. True or False: Vygotsky emphasized the importance of culture and social experience, as well as the role of heredity and brain growth in cognitive change. (25)

14. Bronfenbrenner's ecological systems theory views the person as developing within a ______________________ affected by ______________________. (25)

15. Match each level of ecological systems theory with the appropriate description or example: (25-26)

____	Relationship between the child's home and school	1. Exosystem
____	The influence of cultural values	2. Microsystem
____	The parent's workplace	3. Mesosystem
____	The child's interaction with parents	4. Macrosystem

16. Shifts in contexts called ________________________________ take place throughout life and often represent turning points in development. List one or more examples: ____. (26)

17. In ecological systems theory, development is neither controlled by ________________________ nor by __________________________, but rather by a dynamic interaction between the two. (26)

Comparing and Evaluating Theories

1. Identify the stand that each of the following modern theories take on the view of the developing person, the course of development, and the determinants of development: (27-28)

Theory	Continuous vs. Discontinuous	One Course of Development vs. Many	Nature vs. Nurture
Psychoanalytic theory	____________	____________	____________
Behaviorism and social learning theory	____________	____________	____________
Piaget's cognitive-developmental theory	____________	____________	____________
Information processing	____________	____________	____________
Ethology	____________	____________	____________
Vygotsky's sociocultural perspective	____________	____________	____________
Ecological systems theory	____________	____________	____________
Lifespan Perspective	____________	____________	____________

Studying Development

1. Research usually begins with a ____________________, or a prediction about behavior drawn from a theory. (27)

2. List one strength and two limitations of naturalistic observation. (31)
Strength: ____________________
Limitation: ____________________
Limitation: ____________________

3. List one strength and one limitation of structured observation. (31)
Strength: ____________________
Limitation: ____________________

4. List two strengths and two limitations of the clinical interview. (29)
Strength: ____________________
Strength: ____________________
Limitation: ____________________
Limitation: ____________________

5. List two strengths and two limitations of structured interviews. (30-31)
Strength: ____________________
Strength: ____________________
Limitation: ____________________
Limitation: ____________________

6. List one strength and two limitations of the clinical, or case study, method. (30-31)
Strength: ____________________
Limitation: ____________________
Limitation: ____________________

7. Describe the goals of ethnographic research. How are these goals achieved? (30-31)

____

____

____

8. List one strength and two limitations of ethnographic research. (30-31)
Strength: ____________________
Limitation: ____________________
Limitation: ____________________

9. Define *correlational design*, and give an example of a research question that it is suited to answer. (31-32)
Definition: ____________________
Example: ____________________

10. True or False: A correlational design does not permit researchers to determine cause-and-effect relationships between variables. (32)

11. What is the function of a *correlation coefficient?* (32)

____

____

12. Interpret the meaning of the following two correlations: +.70 between intelligence and school achievement; –.45 between warm, consistent parenting and delinquency. (32-33)
+.70: ____________________
–.45: ____________________

13. What is the unique feature of an *experimental* as opposed to a correlational design? (34)

____

14. The ____________ variable is the one which is expected to cause changes in another variable. The ____________ variable is the one which is expected to be influenced. (34)

15. One procedure that helps control for unknown characteristics of participants that could reduce the accuracy of experimental findings is ____________ ____________. (34)

16. Why might a researcher choose to conduct a field experiment or a natural experiment as opposed to a laboratory experiment? (34)
Field experiment: ____________________
____
Natural experiment: ____________________
____

17. True or False: Experiments conducted in the field are typically just as well controlled and rigorous as laboratory experiments. (35)

18. Describe two strengths and three problems in conducting longitudinal research. (35-37)
Strength: ____________________
Strength: ____________________
Problem: ____________________
Problem: ____________________
Problem: ____________________

19. Describe the cross-sectional design. (37-38)

____
____

20. List one strength and two problems in conducting cross-sectional research. (35, 38)
Strength: ____________________
Problem: ____________________
Problem: ____________________

21. Describe the longitudinal-sequential design. (38)

____
____

22. List two advantages of the longitudinal-sequential design. (35, 38)
A. ____________________
B. ____________________

23. True or False: Transition to a new country has a negative impact on the psychological well-being of children of immigrant parents. (32)

24. Parental education and income can influence how successfully immigrant children adapt. List three other factors that can influence adaptation. (32)
A. ______________________________
B. ______________________________
C. ______________________________

25. How did the adjustment of adolescent girls in economically deprived homes during the Great Depression of the 1930s differ from that of adolescent boys? Explain. (36-37)

____
____
____

26. The Guidance study of war veterans showed that the occurrence of ______________________ in 1939 reversed the early negative impact of the Great Depression. (36-37)

Ethics in Lifespan Research

1. The ultimate responsibility for the ethical integrity of research lies with ______________________. However, if special committees existing in colleges, universities, and other institutions identify unjustified negative implications for the welfare of participants, then priority is always given to ______________________. (39)

2. When children are 7 years of age and older, informed consent should be obtained from both ____________________ and ____________________. (40)

3. True or False: Like children, most older adults require more than the usual informed consent procedures. (40)

4. Debriefing should occur when researchers use ______________________ and/or ______________________. (40)

ASK YOURSELF . . .

What stance does the lifespan perspective take on the issue of one course of development or many? How about nature versus nurture and stability versus change? Explain. (7-9)
One course of development or many? ______________________________
____
____
Nature or nurture? ______________________________
____
____
Stability versus change? ______________________________
____
____

List as many examples as you can find of age-graded, history-graded, and nonnormative influences in Sofie's story at the beginning of this chapter. (see text pages 4, 9)
Age-graded:____

____

____

History-graded: ____

____

____

Nonnormative:____

____

____

Suppose we could arrange a debate between John Locke and Jean Jacques Rousseau on the nature-nuture controversy. Summarize the argument that each of these historical figures is likely to present. (see text page 13-14)
John Locke:____

____

Jean Jacques Rousseau:_______________________________________

____

____

Explain how central assumptions of the lifespan perspective are reflected in Teten's and Carus's philosophies of adulthood and aging. (see text page 14)

____

____

____

____

A 4-year-old becomes frightened of the dark and refuses to go to sleep at night. How would a psychoanalyst and a behaviorist differ in their view of how this problem developed? (see text pages 16-19)

____

____

____

What biological concept is emphasized in Piaget's cognitive-developmental approach? From which nineteenth-century theory did Piaget borrow this idea? (see text page 20)
A. ______________________________ B. ______________________________

Refer to Erikson's psychosocial stage of middle adulthood, generativity versus stagnation, described in Table 1.3 on page 18. Did Sofie, in the story at the beginning of this chapter, resolve this psychological conflict positively or negatively? Explain. (see text page 4)

____

____

____

____

____

What shortcoming of the information-processing approach is a strength of ethology, ecological systems theory, and Vygotsky's sociocultural theory? (see text pages 23-26)

____

What features of Vygotsky's sociocultural theory make it very different from Piaget's theory? (see text page 24)

Is Bronfenbrenner's ecological systems theory compatible with assumptions of the lifespan perspective: development as lifelong, as multidirectional, as highly plastic, and embedded in multiple contexts? Explain. (see text pages 9-11, 25-26)

The lifespan perspective points out that nonnormative events can affect development in powerful ways. Reread the description of nonnormative influences on page 12. Which research method would be mostly likely to tap them? (see text pages 28-31)

A researcher is interested in how elders experience daily life in different cultures. Which method should she use? Explain. (see text pages 30-31)

A researcher compares older adults with chronic heart disease to those who are free of major health problems and finds that the first group scores much lower on mental tests. Should the researcher conclude that heart disease causes declines in intellectual functioning in late adulthood? Are other interpretations possible? Explain. (see text pages 31-33)

A researcher wants to find out if children enrolled in child-care centers during the first few years of life achieve as well in elementary school as those reared fully at home by parents. Which developmental design, longitudinal or cross-sectional, is appropriate for answering this question? Why? (see text pages 35-38)

SUGGESTED READINGS

Cavanaugh, J.C., & Whitbourne, S.K. (1999). *Gerontology: an interdisciplinary perspective.* New York: Oxford University Press. Provides a thorough, broad-based survey of current knowledge on aging. Includes chapters on theory, research methods, physical changes, physical and mental health, cognition, self and personality, work and retirement, and dying and bereavement.

Goldhaber, D.E. (2000). *Theories of human development: integrative perspectives.* Mountain View, CA: Mayfield. Presents an extensive summary and critique of historical and current developmental theories, including Bandura's social–cognitive theory, Piaget's cognitive–developmental theory, Freud's and Erikson's psychoanalytic theories, social learning theory, information processing, ethology, Vygotsky's sociocultural theory, and the lifespan perspective.

Miller, S. A. (1998). *Developmental research methods* (2nd ed.). Englewood Cliffs, NJ: Prentice-Hall. Presents a summary of research designs in the field of child development.

Parke, R. D., Ornstein, P. A., Rieser, J. J., & Zahn-Waxler, C. (Eds.). (1994). *A century of developmental psychology*. Washington, DC: American Psychological Association. Provides a collection of chapters detailing the life histories and main tenets of the most influential in child development. Begins with the founding years of the field and concludes with current reflections on the status of developmental theory and research.

Pellegrini, A. D., & Bjorklund, D. (1998). *Applied child study: A developmental approach* (3rd ed.). Mahwah, NJ: Erlbaum. Presents theories, principles, and methods for studying children in the varied contexts in which they live and develop.

PUZZLE 1.1 TERM REVIEW

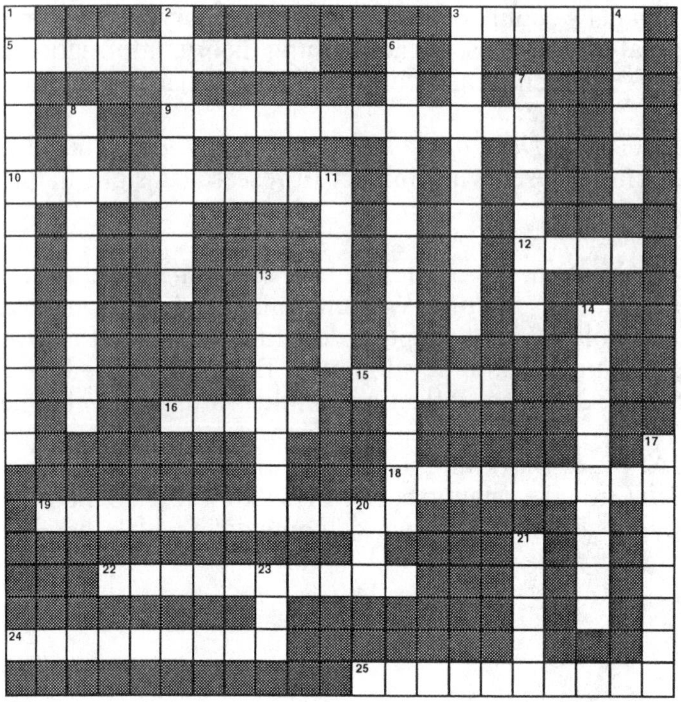

Across

3 Another term used for history-graded influences
5 A longitudinal-_____ design is composed of a sequence of samples that are each followed for a number of years.
9 A(n) _____ design permits inferences about cause and effect.
10 During a structured _______ in the laboratory, an investigator sets up a situation that evokes the behavior of interest.
12 Investigators use a _____ study when they desire an in-depth understanding of a single individual.
15 _____-cultural theory focuses on how culture is transmitted to the next generation.
16 _____ savage: view of child as possessing an innate plan for healthy growth
18 _____ period: optimal time for certain capacities to emerge
19 __________: Bronfenbrenner's temporal dimension
22 Genetically determined course of action
24 _____ approach: measures of behavior are taken on large numbers of individuals
25 Information ______ views the mind as a symbol-manipulating system through which information flows.

Down

1 _____ perspectives assume that children move through stages in which they confront psychological conflicts.
2 During a structured ____, individuals are all asked the same questions in the same way.
3 A process that involves gradually adding on more to the same types of skills that were there to begin with
4 An orderly, integrated set of statements that describes, explains, and predicts behavior
6 Medieval view of the child as a miniature adult
7 Piaget used a _____ interview to study children's thinking.
8 Bronfenbrenner's second level of the environment
11 View that genetic factors are most important in human development
13 Theory concerned with the adaptive value of behavior
14 The ability to adapt effectively in the face of adversity
17 Social _____ theory is based on behaviorism and emphasizes the impact of modeling
20 _____-system: Bronfenbrenner's level that refers to social settings that do not contain the developing person
21 ____-sectional design: groups of people differing in age are studied at the same point in time
23 _____-graded influences: those strongly related to age

PUZZLE 1.2 TERM REVIEW

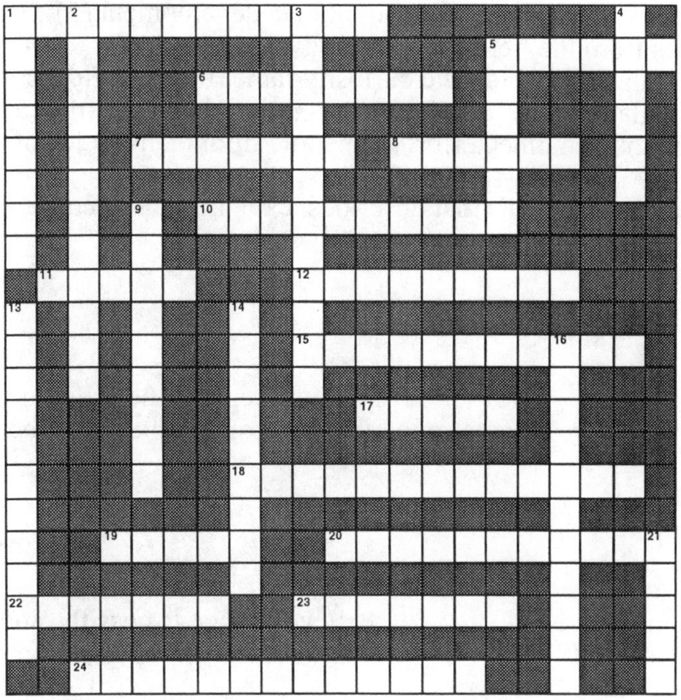

Across

1 ____ design: a group of participants is studied repeatedly at different ages
5 Psycho-_____ theory: Erikson views the ego as a positive force in development
6. Unique combinations of biological and environmental circumstances individuals experience
7 _____-graded influences: those unique to a particular era
8 _____ rasa: blank slate
10 _____ systems theory views the person as developing with in a complex system of relationships at multiple levels.
11 Qualitative change characterizing a particular time period of development
12 _____ vs. change: disagreement about the stability of individual differences
15 Variable manipulated by researchers
17 _____-system: Bronfenbrenner's innermost level of the environment
18 View of development that assumes the existence of stages important in human development
19 _____ development: field devoted to understanding constancy and change throughout the lifespan
20 Approach that focuses on stimuli and and responses
22 View that environmental factors are most important in human development
23 _____ behavior analysis: procedures that combine conditioning and modeling to eliminate undesirable behavior and increase socially acceptable responses
24 Design in which researchers gather information on an already existing group

Down

1 ____ perspective: assumes that development is multidirectional and highly plastic
2 _____ influences happen to just one or a few people
3 Observations conducted in the "field"
4 _____ assignment increases the chances that characteristics of participants do not interfere with the accuracy of experimental findings.
5 Psycho-_____ theory: Freud focuses on early sexual and aggressive drives
9 _____-developmental theory: children actively construct knowledge through direct interaction with the external world
13 A number describing how two variables are related is called a correlation _____.
14 Variable expected to be influenced by manipulations
16 A descriptive, qualitative research method borrowed from the field of anthropology
21 _____-system: Bronfenbrenner's outermost level of the environment

SELF–TEST

1. Our vast storehouse of information about human development (5)
   a. largely comes from the field of psychology.
   b. has been stimulated by scientific curiosity rather than practical concerns.
   c. has been stimulated by practical concerns rather than scientific curiosity.
   d. grew through the combined efforts of people from many fields of study.

2. An integrated set of statements that describes, explains, and predicts behavior is a (5)
   a. strategy.
   b. theory.
   c. hypothesis.
   d. design.

3. The ________________ view of development regards the difference between the immature and the mature being as one of amount or complexity of behavior. (6-7)
   a. discontinuous
   b. continuous
   c. pathwise
   d. stable

4. Theorists who emphasize __________ believe change is possilbe and likely if new experiences support it. (7)
   a. early experience
   b. heredity
   c. stability
   d. plasticity

5. According to the lifespan perspective, which of the following age periods is of supreme importance in development? (9)
   a. early childhood
   b. adolescence
   c. middle adulthood
   d. all age periods have equally powerful effects

6. In medieval times (13)
   a. the Church defended the innocence of children.
   b. children were unprotected by laws.
   c. there were no theories about the uniqueness of children.
   d. toys were designed to amuse children.

7. The idea that children are neither innately good nor evil, but are shaped by their experiences is known as (13)
   a. preformationism
   b. original sin
   c. tabula rasa
   d. noble savage

8. Rousseau felt that parents should (14)
   a. mold their children's behavior.
   b. be receptive to their child's needs.
   c. use harsh parenting strategies.
   d. ignore their children.

9. Tetens was ahead of his time in recognizing that (14)
   a. intellectual declines in old age may reflect hidden gains.
   b. events occurring during childhood have little influence on adult development.
   c. aging is a process of progression in all domains.
   d. adult's thinking is qualitatively different than children's.

10. An important concept emphasized by Darwin which eventually found its way into important mid-twentieth century theories is that of (14-15)
    a. natural, instinctive mother-infant bonding.
    b. the tendency of animal species to become extinct over time.
    c. the adaptive value of physical characteristics and behavior.
    d. the lack of variation within each species which enables it to survive.

11. Which of the following is NOT true of the normative approach to child study? (15)
    a. It is a subjective, observational account of one child's behaviors.
    b. It uses quantitative measurements of a large number of children.
    c. It uses objective measurement techniques including tests and questionnaires.
    d. It was founded by G. Stanley Hall.

12. The success of the Stanford-Binet Intelligence Scale led to (15)
    a. new research addressing the continuous versus discontinuous controversy.
    b. increased attention to individual differences in development.
    c. the comparison of scores of people with different backgrounds.
    d. both b and c.

13. According to Freud's theory, the id (16)
    a. develops through experience.
    b. contains the values of society.
    c. is inherited and present at birth.
    d. mediates between the ego and superego.

14. An approach that emphasizes the role of modeling, or observational learning, in the development of behavior is known as (18-20)
    a. behaviorism.
    b. cognitive-developmental theory.
    c. social learning theory.
    d. operant conditioning theory.

15. A classroom based on Piaget's cognitive-developmental theory would emphasize (20-21)
    a. formal teaching by adults.
    b. modeling by adults and peers.
    c. structured, age-appropriate activities.
    d. discovery learning and direct contact with the environment.

16. Which of the following is NOT a limitation of the information-processing approach? (22-23)
    a. It tells us little about the processes of creativity and imagination.
    b. It has little practical utility.
    c. It tells us little about the links between cognition and other areas of development, such as motivation and emotion.
    d. It isolates children from real-life learning situations.

17. Ethology led to the concept(s) of (23-24)
    a. the four levels of the environment.
    b. imprinting and the critical period.
    c. culturally specific practices.
    d. personal standards and self-efficacy.

18. According to ____________, cognitive development is a socially mediated process. (24)
    a. Piaget
    b. Vygotsky
    c. Bronfenbrenner
    d. Hull

19. Ecological systems theory was developed by (25)
    a. Urie Bronfenbrenner.
    b. Jean Piaget.
    c. Albert Bandura.
    d. B. F. Skinner.

20. To understand human development at the level of the microsystem, one must keep in mind that all relationships are (26)
    a. universal.
    b. unidirectional.
    c. predetermined.
    d. bidirectional.

21. A major limitation of naturalistic observation is that (28-29)
    a. not all participants have the same opportunity to display the behavior of interest.
    b. investigators are not able to quantify their observations.
    c. it is more open to the influence of bias than are other methods.
    d. the resulting data are of little use since underlying thoughts and motives cannot be observed.

22. A number that describes how two variables are related is called a (32)
    a. correlation coefficient.
    b. summation.
    c. standard deviation.
    d. variance.

23. The independent variable is (34)
    a. affected by the dependent variable.
    b. manipulated by the experimenter.
    c. a subject characteristic, such as age.
    d. eliminated through the use of random assignment.

24. Imagine that a researcher finds that 5-year-olds in the 1950s acted differently than do the 5-year-olds of today. Which of the following poses a particular threat to the accuracy of such a finding? (38)
    a. biased sampling
    b. investigator bias
    c. cohort effects
    d. age effects

25. In which of the following cases would extra measures be necessary to ensure the appropriateness of informed consent procedures? (40)
    a. Children over the age of 7 participating in a study that does not involve deception
    b. Elderly adults from residential neighborhoods participating in a study concerning family relationships
    c. Both of the above cases require extra informed consent measures.
    d. Neither of the above cases require extra informed consent measures.

CHAPTER 2

BIOLOGICAL AND ENVIRONMENTAL FOUNDATIONS

BRIEF CHAPTER SUMMARY

This chapter takes a close look at the foundations of development: heredity and environment. With the joining of sperm and ovum at conception, chromosomes containing genetic information from each parent combine to determine characteristics that make us human and contribute to individual differences in appearance and behavior. Several different patterns of inheritance are involved, ensuring that each individual will be unique. Serious developmental problems are often caused by the inheritance of harmful recessive genes and chromosomal abnormalities. Fortunately, genetic counseling and prenatal diagnosis help people at risk for transmitting hereditary disorders assess their chances of giving birth to a healthy baby.

Environmental influences on development are equally as complex as hereditary factors, and play a powerful role throughout the lifespan. The family has an especially powerful impact, operating as a complex social system in which members exert direct, indirect, and third-party influences on one another. Family functioning and individual well-being are influenced considerably by social class differences in child-rearing practices as well as poverty and homelessness. The quality of community life in neighborhoods, schools, towns, and cities, also affects children's and adults' development. Cultural values as well as laws and government programs shape experiences in all of these contexts.

Researchers view the relationship between genetic and environmental factors in different ways. Some believe that it is useful and possible to answer the question of how much each contributes to behavior. Others think that the effects of heredity and environment cannot be clearly separated. They want to discover how these two major determinants of development work together in a complex, dynamic interplay.

LEARNING OBJECTIVES

After reading this chapter, you should be able to:

2.1. Describe the relationship between phenotype and genotype. (46)

2.2. Describe the structure of the DNA molecule, and explain the process of mitosis. (47)

2.3. Explain the process of meiosis. (47)

2.4. Explain how the sex of the new individual is determined. (47-48)

2.5. Identify two types of twins, and explain how each is created. (48)

2.6. Describe basic patterns of genetic inheritance, and indicate how harmful genes are created. (48-53)

2.7. Describe Down syndrome and common abnormalities of the sex chromosomes. (53-54)

2.8. Discuss reproductive options available to prospective parents and the controversies related to them. (60)

2.9. Describe family functioning from the perspective of ecological systems theory, including direct and indirect influences and the family as a dynamic, changing system. (62-63)

2.10. Discuss the impact of socioeconomic status and poverty on family functioning. (63-65)

2.11. Summarize the roles of neighborhoods, towns, and cities in the lives of children and adults. (65-66)

2.12. Explain how cultural values and political and economic conditions affect human development. (66-70)

2.13. Describe and evaluate methods researchers use to determine "how much" heredity and environment influence complex human characteristics. (71-73)

2.14. Describe concepts that indicate "how" heredity and environment work together to influence complex human characteristics. (73-75)

STUDY QUESTIONS

1. __________________ are directly observable characteristics which depend in part on the individual's __________________, the complex blend of genetic information transmitted from one generation to the next. (46)

Genetic Foundations

1. Chromosomes are made up of a chemical substance called __________. It looks like a ________________________ and is composed of segments called ______________. (47)

2. The process through which DNA duplicates itself so that each new body cell contains the same number of chromosomes is called ____________________. (47)

3. The sex cells are referred to as _________________. They are unique in that they contain only _____________ chromosomes, ________________ as many as a regular body cell. (47)

4. The process of ___________________ helps explain why siblings differ from each other. (47)

5. True or False: The female is born with all of her ova already present in her ovaries. (47)

6. The 22 matching pairs of chromosomes are called _______________________. The remaining pair is made up of the _________________________________. (48)

7. What determines whether the new organism will be male or female? (48)

____

8. Match each of the following terms with the appropriate description: (48)

______	The frequency of this type of multiple birth is about 3 out of every 1000 births	1. Fraternal or dizygotic twins
______	The most common type of multiple birth	
______	Genetically no more alike than ordinary siblings	2. Identical or monozygotic twins
______	Genetically identical	

9. Children of single births are often healthier and develop more rapidly than twins in the early years because twins are more often born __________ and get __________ attention than the average single infant. (48)

10. If the genes from both parents are alike, the child is __________________ for a particular trait and will display it. If the genes are different, the child is __________________ for that trait. (49)

11. Give an example of a characteristic representing dominant-recessive inheritance. ____________________________________. (49)

12. Phenylketonuria, or PKU, follows the rules of ______________________ __________________ inheritance and affects the way the body breaks down ________________________. The treatment involves __________________ ____. (49-50)

13. Why are serious diseases only rarely due to dominant genes? Explain why Huntington disease is a dominant disorder which has survived. (50)

A.____
____

B.____

14. *Codominance* is a pattern of inheritance in which ____________________ influence the individual's phenotype. Under what conditions is the sickle cell trait expressed by heterozygous individuals? ____________________________ ____. (50-51)

15. True or False: Males are more likely to be affected by X-linked inheritance than are females. (51-52)

16. Name two X-linked conditions or disorders. (51-52)

A. ______________________________ B. ______________________________

17. Describe *genetic imprinting.* (52-53)

____
____

18. List three disorders that appear to be expressed through genetic imprinting on the autosomes. (51, 53)

A. ______________________________ B. ______________________________

C. ______________________________

19. Name two developmental problems that are associated with fragile X syndrome. (53)

A. ______________________________ B. ______________________________

20. Mutation is ____ and has been known to be caused by __________________________. (53)

21. True or False: In *polygenic inheritance,* one gene determines the characteristic in question. (53)

22. List the behavioral and physical characteristics of Down syndrome. (53-54)
Behavioral: ____
____
Physical: ____
____

23. True or False: The occurrence of Down syndrome rises with maternal age. (53-54)

24. In contrast to autosomal disorders, sex chromosomal disorders often go undetected until ______________________. (54)

25. Geneticists do not know why adding to or subtracting from the usual number of X chromosomes results in very specific ______________________ deficits. (54)

Reproductive Choices

1. If a family history of hereditary problems exists, a couple may seek help through ________________________________. A pedigree is created which is used to ____. (55)

2. True or False: Nearly all states have legal guidelines for donor insemination and in vitro fertilization procedures and require doctors to keep records of donor characteristics. (56)

3. Describe some risks associated with surrogate motherhood. (56)
____
____
____

4. Research on ____ is important for weighing the pros and cons of new reproductive technologies. (57)

5. ____________________ is available for women who have reason to be especially concerned about fetal defects and diseases. (55, 58)

6. List five prenatal diagnostic methods. (55)
A. ______________________ B. ______________________
C. ______________________ D. ______________________
E. ______________________

7. Describe one suggestion for helping parents make informed decisions about fetal surgery. (58)
____
____

8. Advances in ______________________ offer hope for correcting hereditary defects. (58)

9. List three goals of the Human Genome Project. (60-61)
A. ______________________
B. ______________________
C. ______________________

10. True or False: Research tools used by the Human Genome Project only allow researchers to understand disorders due to single genes. (60-61)

11. List four gene-based treatments being developed by the Human Genome Project. (60-61)
A. ______________________
B. ______________________
C. ______________________
D. ______________________

12. The risk of adoption failure is greatest for ______________________ ________ and ______________________, but it is not high. (58-59)

13. List three possible reasons that adopted children have more emotional and learning difficulties than do nonadopted children. (59)
A. ______________________
B. ______________________
C. ______________________

14. True or False: Most adoptees appear well adjusted as adults. (59)

Environmental Contexts for Development

1. Gratifying family ties predict ______________________, whereas alienation from the family is often associated with ______________________. (62)

2. ______________ family influences are those in which the behavior of one person helps sustain a form of interaction in the other that either promotes or undermines psychological well-being. (62)

3. ________________ family influences are interactions between any two members that are affected by others present in the setting. (62)

4. Explain how important events and historical time period contribute to the dynamic, ever-changing nature of the family. (63)

____
____
____

5. Briefly describe socioeconomic status differences in the timing and duration of family life cycle phases, values and expectations, and family interaction. (63)
Timing/duration of phases:____

____

Values and expectations:____

____

Family interaction:____

____

6. Name two general influences which help explain the differences listed above. (63)
A. ____
B. ____

7. What two groups are hit hardest by poverty? (64)
A. ______________________________ B. ________________________________

8. Describe the difficulties from which homeless children suffer. (64-65)

____

____

9. When ties between family and community are ________________, family stress and child adjustment problems are reduced. (65)

10. List two ways in which neighborhoods can affect children's development and two ways in which they affect well-being in adulthood. (65)
Two ways in childhood: ____

____

Two ways in adulthood:____

____

11. True or False: In small towns, connections between settings that influence children's lives are common. (66)

12. Adults in small towns are more likely to ___________________________________ ____________________. (66)

13. Why is community life especially undermined in high-rise urban housing projects? (66)

____

____

____

____

14. What are subcultures? (66)

____

____

15. One important way in which African American families often differ from other American families is the black tradition of _________________________________ _________________. (66-67)

16. By providing _______________________ and sharing _____________________ ____________________________, the African-American extended family helps reduce the stress of poverty and single parenthood. (67)

17. African-American elderly report a ________________ degree of satisfaction from family life. (67)

18. Define *collectivist* and *individualist societies.* Does the United States emphasize collectivism or individualism? (68)
Collectivist societies: ____

____

Individualist societies: ____

____

The United States? ____

19. American public policies safeguarding children and youths have (lagged behind/surpassed) those for the elderly. Both sets of policies have emerged (slower/faster) in the United States than in other Western industrialized nations. (68)

20. List three reasons why the United States has not yet created conditions that protect the development of children. (69)
A. ____
B. ____
C. ____

21. Today, spending for the aged accounts for _______ percent of all federal expenditures—5 times what was spent 30 years ago. However, American policies on aging have been criticized for neglecting ________________________
______________________________. (69)

22. How has the number of aging poor changed since 1960, and how are the elderly in the United States faring in comparison to those in Australia, Canada, and Western Europe? (69)
A. ____
B. ____

23. List the activities of the Children's Defense Fund. (70)

____
____
____

24. List the activities of the American Association of Retired Persons. (70)

____
____
____

Understanding the Relationship Between Heredity and Environment

1. Scientists must study the impact of genes on ____________________ traits indirectly. (71)

2. Heritability estimates and concordance rates are used to answer the question of ______________ heredity and environment contribute to complex human characteristics. (71-72)

3. Heritability estimates are obtained from _____________ studies. The most common type compares _____________________ with _______________________ twins. (71)

4. The heritability estimate for intelligence in child and adolescent twin samples is about ________, which indicates that _______________ of individual differences in intelligence can be explained by genetic makeup. (71-72)

5. A concordance rate of 0 means that if one twin has a trait the other one (always/never) has it. A concordance rate of 100 means that if one twin has a trait, the other one (always/never) has it. (72)

6. List three limitations of heritability estimates and concordance rates. (72-73)
A. ____
B. ____
C. ____

7. Describe *range of reaction* and *canalization.* (73-74)
A. Range of reaction: ____
____
____
B. Canalization: ____
____
____

8. Match the following types of genetic-environmental correlations with their appropriate descriptors. (74-75)

____	Children increasingly seek out environments that fit their genetic tendencies (called *niche-picking*).	1. passive correlation
____	A child's style of responding influences others' responses, which then strengthen the child's original style.	2. evocative correlation
____	Parents provide an environment consistent with their own heredity.	3. active correlation

9. True or False: Partially as a result of niche-picking, identical twins become more similar and fraternal twins and adopted siblings become less similar in intelligence with age. (75)

10. The success of any attempt to improve development depends on what three factors? (75)
A. ____
B. ____
C. ____

ASK YOURSELF . . .

Gilbert and Jan are planning to have children. Gilbert's genetic makeup is homozygous for dark hair; Jan's is homozygous for blond hair. What color is Gilbert's hair? How about Jan's? What proportion of their children are likely to be dark-haired? (see text pages 48-49)
Gilbert's hair color: ____________________ Jan's hair color: ____________________
Proportion of dark-haired children: ____

Ashley and Harold both carry the defective gene for fragile X syndrome. Explain why Ashley's child inherited the disorder but Harold's did not. (see text pages 51-53)

____

____

Cite a recessive disorder that leads to profound disability but for which early environmental intervention can change the life course. Explain how heredity and environment work together in this example. (see text pages 49-50)
Recessive disorder:____
How heredity and environment work together:__________________________________

____

____

A woman over 35 has just learned that she is pregnant. Although she would like to find out as soon as possible whether her child has a chromosomal disorder, she wants to minimize the risk of injury to the developing baby. Which prenatal diagnostic method is she likely to choose? (see text page 55, 58)

____

Describe the ethical pros and cons of fetal surgery, surrogate motherhood, and postmenopausal-assisted childbearing. (see text pages 56-57)

____

____

____

____

How does research on adoption reveal resiliency at several periods of the life course: childhood, adolescence, and early adulthood? Of the factors related to resiliency (see Chapter 1, page 10), which appears to be central in positive outcomes for adoptees? (see text pages 58-60)
Childhood:____
Adolescence:____
Early adulthood: ____
Factors central in positive outcomes:____

____

On one of your trips to the local shopping center, you see a father getting very angry at his young son. Using ecological systems theory, list as many factors as you can that might account for the father's behavior. (see text pages 62-63)

____

____

____

____

Links between family and community foster development throughout the lifespan. Provide examples and research findings from our discussion that support this idea. (see text pages 65-66)

____
____
____
____

Check your local newspaper and one or two national news magazines to see how often articles appear on the condition of children, families, and the aged. Why is it important for researchers to communicate with the general public about the well-being of these sectors of the population? (see text pages 68-70)

____
____
____

A researcher wants to know whether alcoholism is influenced by genetic factors. Which method of inferring the importance of heredity in complex human characteristics could help answer this question? (see text page 72)

____

One possible reason for the increase in heritability of intelligence in adulthood is that adults exert more control over their intellectual experiences than do children. What type of genetic-environmental correlation is involved in this explanation? What light does it shed on the question of whether heritability estimates yield "pure" measures of genetic influences on human traits? (see text pages 71-73)

Type of genetic-environmental correlation: ____________________

Do heritability estimates yield "pure" measures of genetic influences on traits?

____
____

SUGGESTED READINGS

Clarke, A. (Ed.) (1994). *Genetic counseling: Practice and principles.* New York: Routledge. Examines social and ethical issues raised in genetic counseling. Chapters are written by experts from a variety of fields, including medicine, social science, philosophy, and law. The result is a book that looks at genetic counseling from many perspectives.

Elder, K., & Dale, B. (2000). *In vitro fertilization.* New York: Cambridge University Press. Describes the most recent additions to the range of current assisted reproductive technology (ART) clinical treatments, including the use of testicular and epididymal sperm, blastocyst stage transfer, and new perspectives on cryobiology and cryopreservation techniques.

Gottlieb, G. (1997). *Synthesizing nature–nurture: Prenatal roots of instinctive behavior.* Mahwah, NJ: Erlbaum. Provides a nontechnical account of theoretical and experimental work conducted over the course of 35 years that resulted in a framework capable of integrating causal influences at the genetic, neural, behavioral, and ecological levels of development.

Marshall, E. L. (1997, paperback). *The Human Genome Project: Cracking the code within us*. Danbury, CT: Franklin Watts. Describes the Human Genome Project, which is aimed at deciphering the chemical makeup of human genetic material. Addresses the controversies surrounding the field of genetic manipulation, but emphasizes the benefits of the genetic map for the future prevention and treatment of genetic disorders.

Rickel, A. U., & Becker, E. (1997). *Keeping children from harm's way: How national policy affects psychological development*. Washington DC: American Psychological Association. Discusses, from the perspective of these Congressional Fellows, the 1990s era of social policy and its denial of family needs, preventable risk factors affecting children's well-being, successful intervention and prevention programs, special issues for adolescents, and transgenerational life outcomes.

PUZZLE 2.1 TERM REVIEW

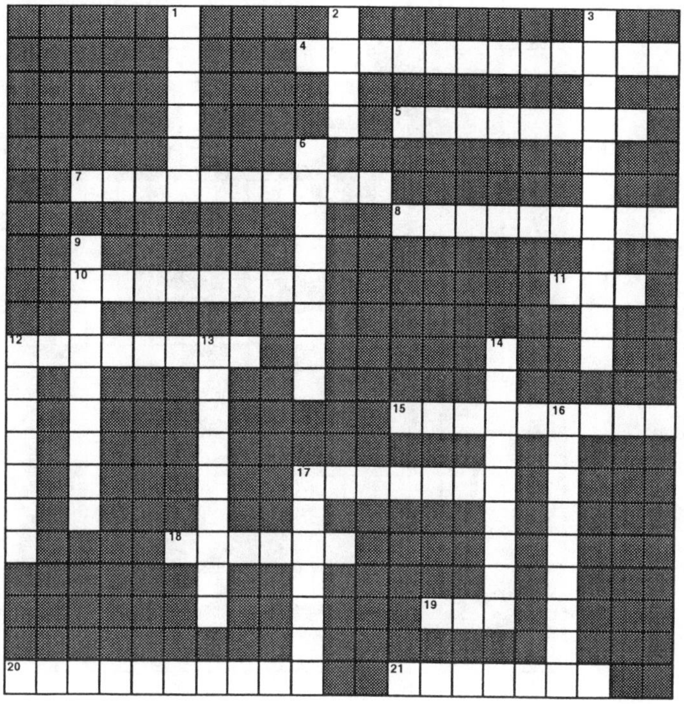

Across

4 Having differing genes at the same place on a pair of chromosomes
5 Range of ____: each person's genetically determined response to a range of environmental conditions
7 A group with beliefs and customs differing from those of the larger culture
8 Influenced by many genes
10 Gene that determines phenotype in a heterozygous combination
11 Contains the genetic code
12 Genetic makeup of the individual
15 Twins who are genetically no more alike than ordinary siblings
17 Process of cell division through which gametes are formed
18 Cell that results when sperm and ovum unite at conception
19 Measure of family's social position and economic well-being (abbr.)
20 If genes from both parents are alike, the child is _____.
21 A heterozygous individual who can pass a recessive gene to offspring

Down

1 Type of policy involving government laws and programs for improving conditions
2 Segment of a DNA molecule
3 Both genes in a heterozygous combination are expressed
6 Unfavorable genes arise from _____.
9 Twins who have the same genetic makeup
12 Human sperm and ova
13 Physical and behavioral characteristics, determined both genetically and environmentally
14 The 22 matching chromosome pairs in each human cell
16 Gene not expressed in a heterozygous combination
17 Process of cell duplication

PUZZLE 2.2 TERM REVIEW

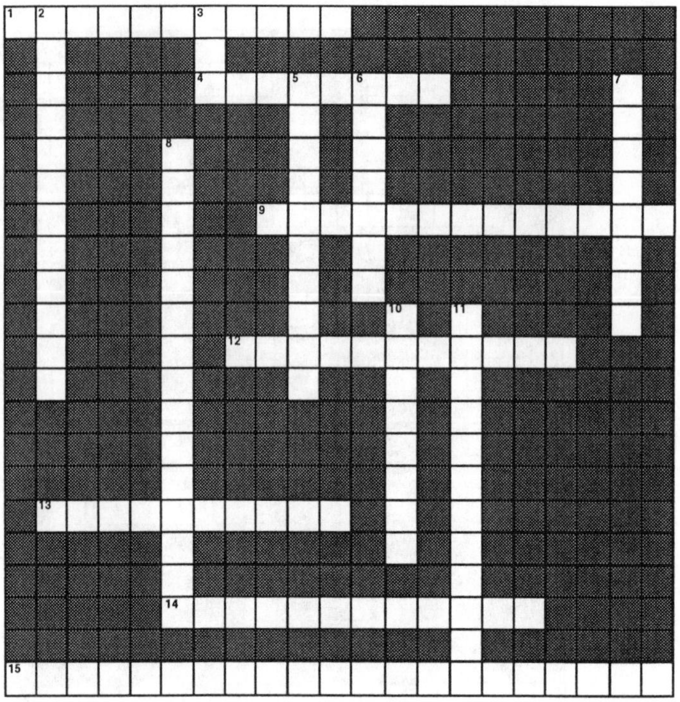

Across

1 Rodlike structures in the cell nucleus that store and transmit information
4 _____ _____ inheritance: a recessive gene is carried on the X chromosome
9 Individuals actively choose environments that complement their heredity (2 words)
12 _____ rate is the percentage of instances in which both members of a twin pair show a trait when it is present in one member.
13 Genetic _____ helps couples assess the likelihood of passing on a hereditary disorder.
14 _____ societies: people stress group over individual goals
15 _____ -_____ correlation is the idea that heredity influences the environments to which children are exposed.

Down

2 _____ estimate: a statistic that measures the extent to which individual differences in complex traits are due to genetics
3 _____ chromosomes: the 23rd pair, XX females and XY in males
5 Genetic _____: genes are chemically marked in such a way that one pair member is activated regardless of its makeup
6 _____ studies compare the characteristics of family members to determine the role of heredity in complex traits.
7 _____ family households: three or more generations live together
8 _____ societies: people are largely concerned with their own personal needs
10 _____ diagnostic methods permit the detection of problems before birth
11 The tendency of heredity to restrict the development of some characteristics to just one or a few outcomes

SELF–TEST

1. Phenotypes are influenced by (46)
   a. only the individual's genotype.
   b. both genotype and environment.
   c. only the individual's environment.
   d. neither genotype nor environment.

2. Chromosomes are made up of (47)
   a. ribonucleic acid.
   b. sequences of potassium.
   c. deoxyribonucleic acid.
   d. genotypes.

3. A unique feature of DNA is that it can (47)
   a. disguise the genetic code.
   b. spontaneously recover from damage.
   c. duplicate itself.
   d. alter the environment.

4. In a heterozygous pairing in which only one gene affects the child's characteristics, the pattern of genetic inheritance is called (49)
   a. dominant–recessive.
   b. codominant.
   c. polygenic.
   d. mutagenic.

5. Luke is homozygous for the recessive trait of blond hair. His wife Sally is homozygous for the dominant trait of dark hair. How many of Luke and Sally's children will be dark-haired? (49)
   a. none
   b. 50 percent
   c. 75 percent
   d. 100 percent

6. Serious diseases are only rarely due to dominant genes because (50)
   a. dominance requires the same gene to be inherited from both parents.
   b. other genes usually alter the effects of dominant genes.
   c. children inheriting the harmful dominant gene would always develop the disorder and seldom live long enough to pass on the trait.
   d. recessive genes usually overtake the harmful dominant gene.

7. The pattern of inheritance in which both genes influence the person's characteristics is called (50-52)
   a. dominant-recessive.
   b. heterodominance.
   c. codominance.
   d. X-linked.

8. Which of the following statements is true? (52)
   a. Learning disabilities are more common among females than males.
   b. There are 106 boys born for every 100 girls.
   c. Mental retardation is more common among females than males.
   d. Behavior disorders are more common among females than males.

9. In which of the following patterns of inheritance are genes chemically marked in such a way that only one pair member is activated, regardless of its makeup? (52-53)
   a. codominance
   b. homozygous inheritance
   c. polygenic inheritance
   d. genetic imprinting

10. Mutations (53)
    a. always occur spontaneously.
    b. are always induced by external factors.
    c. can be either spontaneously or externally induced.
    d. never occur by chance.

11. The most common chromosomal abnormality is (53)
    a. Klinefelter's syndrome.
    b. PKU.
    c. cystic fibrosis.
    d. Down syndrome.

12. ______________________ is a technique involving examination of fluid from the uterus for genetic defects that can be performed by 11 to 14 weeks after conception. (55-58)
    a. Amniocentesis
    b. Chorionic villus sampling
    c. Fetoscopy
    d. In vitro sampling

13. Which of the following is NOT a research tool used by the Human Genome Project? (59-60)
    a. genetic map
    b. environment map
    c. sequence map
    d. physical map

14. Modern theorists view the family as (62)
    a. stable and resistant to change.
    b. dynamic and ever-changing.
    c. a unidirectional system.
    d. immune to "third party" effects.

15. Researchers use all of the following variables to assess socioeconomic status (SES) <u>except</u> (63)
    a. the prestige of one's job.
    b. education.
    c. income.
    d. race.

16. _________________ is/are greatest in neighborhoods where residents feel socially isolated. (65)
    a. Child abuse and neglect
    b. Contact with relatives
    c. Church attendance
    d. Family cohesiveness

17. Compared to large cities, small towns are more likely to foster (66)
    a. feelings of loneliness.
    b. limited social contact.
    c. active involvement in the community.
    d. more visits to museums and concerts.

18. Which of the following has helped to protect African-American children from the harmful effects of poverty? (66-68)
    a. publicly sponsored day care
    b. excellent schools in American inner cities
    c. migration of African Americans to small towns
    d. extended family households

19. The U.S. government hesitates to become involved in family matters because (69)
    a. Americans value self-reliance and privacy.
    e. Americans are in strong agreement on issues of child and family policy.
    f. America is a rich country, and there are ample opportunities for families to help themselves.
    d. America does not view the family unit as critical to society.

20. Laws and government programs designed to improve current conditions are called (68)
    a. legal policies.
    b. extended policies.
    c. public policies.
    d. democratic policies.

21. The status of children in the United States (68-69)
    a. is superior to that of children in most other industrialized nations.
    d. is protected by a strong set of public policies to foster children's development.
    c. lags behind that of many other industrialized nations.
    d. is unlikely to change due to a lack of knowledge and resources on the part of the United States.

22. A statistic that measures the extent to which continuous traits, such as intelligence, can be traced to heredity is called a (71)
    a. heritability estimate.
    b. kinship estimate.
    c. concordance rate.
    d. t-test.

23. Which of the following is true about concordance rates? (72)
    a. Concordance rates are used to study the role of heredity in traits that can be judged to be either present or absent.
    b. If the concordance rate is lower for identical twins than for fraternal twins, heredity is believed to have an important effect on the trait.
    c. A score of 100 means that if one twin has the trait, the other twin will never have the trait.
    d. A score of 100 means that if one twin has the trait, the other twin will always have the trait.

24. Since all normal human babies roll over, sit up, crawl, and walk, one can conclude that infant motor behavior is a strongly _________ trait. (74)
    a. canalized
    b. imprinted
    c. instinctive
    d. encouraged

25. Identical twins reared apart who nevertheless have many psychological traits and lifestyle characteristics in common illustrate a form of genetic-environmental correlation commonly called (75)
    a. range of reaction.
    b. canalization.
    c. niche-picking.
    d. evocative.

CHAPTER 3

PRENATAL DEVELOPMENT, BIRTH, AND THE NEWBORN BABY

BRIEF CHAPTER SUMMARY

At no other time in the lifespan is change as rapid as it is before birth. Prenatal development takes place in three phases: (1) the period of the zygote, (2) the period of the embryo, and (3) the period of the fetus. Various environmental agents and maternal conditions can damage the developing organism, making the prenatal period a vulnerable time. For this reason, prenatal health care is vitally important to ensure the health of mother and baby.

Childbirth takes place in three stages: (1) dilation and effacement of the cervix, (2) delivery of the baby; and (3) birth of the placenta. Production of stress hormones helps the infant withstand the trauma of childbirth, and the baby's physical condition is assessed immediately after birth using the Apgar Scale. Alternatives to traditional hospital childbirth include natural childbirth and delivery in a birth center or at home. Still, childbirth in the United States is often accompanied by a variety of medical interventions. Although they help save the lives of many babies, these procedures can cause problems of their own when used routinely. When birth complications occur, they most often involve preterm and low-birth-weight infants. Fortunately, many of these babies recover from difficulties with the help of supportive home environments.

Infants begin life with a remarkable set of skills for relating to the surrounding world. They display a wide variety of reflexes, and with the exception of vision, their senses are well developed. In the early weeks, babies move in and out of different states of arousal frequently. They first communicate by crying, and with experience, parents become better at interpreting and responding to infants' cries. Tests of newborn behavior allow the assessment of the baby's many capacities. The baby's arrival is exciting, but it brings with it profound changes. Husbands and wives who support each other's needs typically adjust well to the demands of parenthood.

LEARNING OBJECTIVES

After reading this chapter, you should be able to:

3.1. List the three phases of prenatal development, and describe major milestones of each. (80-86)

3.2. Define the term *teratogen*, and summarize factors that affect the impact of teratogens. (86-87)

3.3. List agents known or suspected of being teratogens, and discuss evidence supporting the harmful impact of each. (87-93)

3.4. Discuss maternal factors other than teratogens that can affect the developing embryo or fetus. (93-96)

3.5. Describe the three stages of labor. (96-97)

3.6. Discuss the baby's adaptation to labor and delivery, and describe the appearance of the newborn. (98)

3.7. Explain the purpose and main features of the Apgar Scale. (98-99)

3.8. List the advantages and disadvantages of giving birth in a hospital, birth center, and home, and describe any benefits and concerns associated with natural childbirth and home delivery. (99-100)

3.9. Describe circumstances that justify use of fetal monitoring, labor and delivery medication, and cesarean delivery, and explain any risks associated with each. (100-101)

3.10. Discuss risks associated with low birth weight, and cite factors that can help infants who survive a traumatic birth develop. (102-105)

3.11. Name and describe major newborn reflexes, noting their functions and the importance of assessing them. (106-107)

3.12. Describe newborn states of arousal, including characteristics of sleep and crying and ways to soothe a crying newborn. (107-109)

3.13. Describe the newborn baby's responsiveness to touch, taste, smell, sound, and visual stimulation. (110-111)

3.14. Describe Brazelton's Neonatal Behavioral Assessment Scale (NBAS), and explain its usefulness. (111-112)

3.15. Describe typical changes in the family after the birth of a new baby. (112)

STUDY QUESTIONS

Prenatal Development

1. True or False: The ovum is the largest cell in the human body. (80)

2. Sperm live for up to _____ day(s), whereas the ovum survives for only _____ day(s). (81)

3. How long is the period of the zygote? What event signifies its beginning, and what event signifies its end? (82)

A. ______________________________ B. ______________________________

C. ______________________________

4. After implantation, the protective outer layer develops into a membrane called the ________________, which encloses the developing organism in ________________ ________________________________. (82)

5. Describe the function of the placenta and the umbilical cord. (82)
Placenta: ____

____

Umbilical cord: ____

____

6. The period of the embryo lasts from the __________ through the ____ week of pregnancy. During these 6 weeks, the groundwork is laid for all __________________ and ______________________. (83)

7. What will the three layers of the embryonic disk form? (83)
Ectoderm: ____
Mesoderm: ____
Endoderm: ____

8. Briefly summarize the events that take place during the following embryonic time periods: (83)
Last half of first month: ____

____

____

Second month: ____

____

____

9. During the period of the fetus, the developing organism __________________ ______________________. Briefly cite evidence of the organization and connection between organs, muscles, and nervous system which begin during the third month. (83)

____

____

10. By the end of the second trimester, all the brain's ________________ are in place. (Few more/Many more) will be produced in the individual's lifetime. (84)

11. List two behavioral capacities that emerge during the second trimester. (84)
A. ______________________________ B. ______________________________

12. The prenatal age at which the baby can first survive is called the ________ ___________________________. It occurs between _____ and _____ weeks. (84)

13. True or False: Research shows that the fetus can distinguish the tone and rhythm of the mother's voice during the third trimester. (86)

14. During the final three months of pregnancy, the fetus receives __________________ from the mother's blood to protect against illness. (86)

15. Recent research found moderate links between prenatal measures and infant characteristics. Specifically, one recent study found that the pattern of fetal ___________________ in the last few weeks of pregnancy was the best predictor of infant ___________________. (85)

Prenatal Environmental Influences

1. The term *teratogen* refers to ____
____. (86)

2. Describe four factors that affect the impact of teratogens. (86)
A. ____
B. ____
C. ____
D. ____

3. Why is the embryonic period the time during which serious prenatal defects are most likely to occur? (86-87)
____

4. True or False: The effects of teratogens are limited to physical damage. (87)

5. Researchers have found a link between low birth weight and ____________________, ____________________, and ____________________ for both sexes in several countries. (88)

6. High birth weight has been found to be related to ____________________ in women. (88-89)

7. During what period of prenatal development did thalidomide have the greatest damaging effects? What consequences did it have for later development? (87-88)
A. ____
B. ____

8. What effect does the drug DES when taken by mothers during pregnancy have on young women and men later in life? (88)
Women: ____
Men: ____

9. True or False: Aspirin and caffeine intake during pregnancy are not related to developmental problems. (88-89)

10. Match each of the following illegal drugs with associated problems: (89-90)

_____	Infants exposed prenatally show problems with attention and motor development that may or may not improve after infancy	1. Crack
_____	A cheap form of cocaine associated with low birth weight and damage to the central nervous system.	2. Heroin and methadone
_____	Constricts the blood vessels causing a reduction in oxygen delivered to the developing organism	3. Marijuana
_____	Prenatal exposure is related to newborn startles, disturbed sleep, and reduced attention in infancy and childhood.	4. Cocaine

11. True or False: Fathers who use cocaine may contribute to drug-related birth defects because cocaine can "hitchhike" its way to the zygote by attaching to the sperm. (90)

12. The most well-known effect of smoking during pregnancy is ___________________ ____________, but the likelihood of other problems such as _____________________, _____________________, _____________________, _____________________, and ___________________________ also increase. (90-91)

13. Children who display mental retardation, poor attention, over activity, and a distinct set of physical symptoms and were born to mothers who abused alcohol during pregnancy are said to have _________________________. (91)

14. Infants who show some, but not all, of the abnormalities associated with this disorder are said to suffer from _______________________________. (91)

15. True or False: Babies exposed to radiation who appear normal may display problems later in life. (92)

16. Match each of the following environmental pollutants with its effect on development: (92)

____	When absorbed from car exhaust, paint, and other materials used industrially, the pollutant leads to prematurity, low birth weight, brain damage, and physical defects	1. Mercury
____	In the 1950s, children prenatally exposed in Japan were mentally retarded and physically handicapped	2. Lead
____	Women who ate fish contaminated with this substance gave birth to babies with slightly reduced birth weight, smaller heads, and persistent memory and intellectual deficits	3. PCBs

17. Describe the outcomes associated with embryonic and fetal exposure to maternal rubella. (92)

Embryonic: ____

____

Fetal:____

18. When women carrying the AIDS virus become pregnant, ______ of the time they pass the disease to the embryo or fetus. (93)

19. True or False: AIDs progresses more slowly in infants than in older children and adults. (93)

20. True or False: Bacterial and parasitic diseases have little impact on the prenatal organism. (93)

21. Summarize the behavioral and health problems of prenatally malnourished babies. (93)

____

____

____

22. True or False: Supplementation around the time of conception reduces neural tubal defects. (93)

23. What two approaches to intervention with prenatally malnourished babies are effective? (94)
A. ____________________
B. ____________________

24. Each year, __________ to ____________ American infants are born seriously undernourished. (94)

25. List the birth complications and physical defects associated with intense anxiety during pregnancy: (94)
A. ______________ B. ______________
C. ______________ D. ______________
E. ______________ F. ______________

26. During maternal stress, blood flow to parts of the body involved in the defense response is (increased/decreased) and blood flow to the uterus is (increased/decreased). In addition, ____________________ released into the mother's blood stream cross the placenta and increase the fetus's heart rate and activity level. (94)

27. True or False: Social support cannot reduce the risks of maternal emotional stress during pregnancy on infant development. (94)

28. The Rh factor may cause problems for the developing fetus when the mother is Rh __________________ and the father is Rh __________________. (94)

29. Explain why problems resulting from the Rh factor are more likely to affect later-born children. (94)

____

30. True or False: When women without serious health problems are considered, even those in their forties do not experience more prenatal problems than do those in their twenties. (95)

31. Why are infants of teenagers at greater risk of prenatal difficulties? (95)

____
____

32. In women with diabetes, extra sugar in the mother's bloodstream causes the fetus to grow ____________________, making pregnancy and birth complications more common. (95)

33. In women with a condition called toxemia, ____________________ increases sharply during the last half of pregnancy, and the face, hands, and feet swell. If not treated, toxemia can cause ____________________ ____________________. (95)

34. List three common demographic characteristics of women who do not seek prenatal care until late in their pregnancies. (95)
A. ______________ B. ______________
C. ______________

Childbirth

1. The first stage of labor, which lasts an average of ______ to ______ hours with a first baby, includes the process of ________________________________ of the cervix. (97)

2. The second stage of labor lasts about ______ minutes for a first baby. Briefly, what occurs during this stage? ________________________________. (97)

3. The final stage of labor occurs after the baby's birth and involves the delivery of the __________________. (97)

4. Cite three ways in which stress hormones help the baby withstand the trauma of childbirth. (98)
A. ________________________________
B. ________________________________
C. ________________________________

5. The ______________________ is used to quickly assess the infant's physical condition at birth in terms of the following five characteristics: (98-99)
A. __________________ B. __________________ C. __________________
D. __________________ E. __________________

Approaches to Childbirth

1. Freestanding birth centers encourage ______________________________ between parents and baby. However, they offer less ______________________. (99)

2. Summarize the three main parts of a typical natural childbirth program. (99-100)
A.________________________________
____
B.________________________________
____
C.________________________________
____

3. Most home births are handled by ____________________________________ who have degrees in nursing and training in childbirth management. (100)

Medical Interventions

1. ______________________ is a term for a variety of impairments in muscle coordination that result from brain damage around the time of birth. (100)

2. More than anywhere else in the world, childbirth in the United States is a ____ event. (101)

3. Describe the purpose of fetal monitors, and list three reasons that they are controversial. (101)
Purpose:________________________________
____
A.________________________________
B.________________________________
C.________________________________

4. Cite two reasons that routine use of labor and delivery medication can cause problems. (101)
A.____
____
B.____
____

5. Under what conditions is a cesarean delivery clearly justified? Under what conditions is it only sometimes justified? (101)
Clearly justified: ____
____
____
Sometimes justified:
____
____
____

6. True or False: Once a woman has had a cesarean, she must always give birth by cesarean. (101)

7. How can a cesarean affect the adjustment of the newborn baby and the early mother–infant relationship? (101)
____
____
____

Preterm and Low-Birth-Weight Infants

1. ______________________ is the best available predictor of infant survival and healthy development. (102)

2. Why is low birth weight highest among infants born to low-income women? (102)
____
____

3. Distinguish between *preterm* and *small-for-date* babies.
Preterm:____
Small for date:____
Which group has more problems? Explain why. (101)
____
____

4. Compared to full-term infants, preterm babies are less often __________________, ______________________, and __________________ by parents and are at risk for ____________________. (103)

5. Describe the types of stimulation used and the outcomes associated with the use of special infant stimulation in some intensive care nurseries. (103)
A.____
____
B.____
____

6. True or False: Gentle massage of a preterm infant can lead to faster weight gain and, by the end of the first year, increased mental and motor development. (103)

7. " Kangaroo baby care," in which the preterm infant is tucked between the mother's breasts inside of her clothing, is being encouraged in developing countries. What are the benefits of this intervention? (103)

____

____

8. Briefly describe the characteristics of interventions that best facilitate the parenting of preterm babies. (103)

____

____

9. When preterm infants live in ______________________________ households, long-term, intensive intervention is necessary. To sustain developmental gains in these children, intervention is needed well beyond age _____. (103)

10. ____________________________ refers to the number of deaths in the first year of life per 1,000 live births. In the past three decades, this number has (increased/decreased) in the United States. (104)

11. Neonatal morality is the rate of death within the first month of life and accounts for _____ percent of the infant death rate in the United States. List two factors that are largely responsible for neonatal mortality. (104)

A. ____

B. ____

12. List four incentives European nations offer mothers and young children to increase infant survival rates. (104-105)

A. ____

B. ____

C. ____

D. ____

Understanding Birth Complications

1. Findings of the Kauai Study revealed that the most powerful clue to how well children recovered from birth complications was ____________________ ____________________________________. (104-105)

2. True or False: When family lives are stressful and chaotic, even resilient children are unable to recover from birth complications. (104-105)

The Newborn Baby's Capacities

1. Match each reflex with its appropriate description. (106-107)

_____	Prepares infant for voluntary grasping	1. Eye blink
_____	When sole of foot is stroked, toes fan out and curl	2. Tonic neck
_____	Helps infant find the nipple	3. Palmar grasp
_____	Prepares infant for voluntary walking	4. Babinski
_____	Permits feeding	5. Rooting
_____	Infant lies in a "fencing position"	6. Sucking
_____	Protection from strong stimulation	7. Swimming
_____	In our evolutionary past, may have helped infant cling to mother	8. Stepping
_____	Helps infant survive if dropped in water	9. Moro
_____	Protects infants from unpleasant tactile stimulation	10. Withdrawal

2. Some reflexes (such as sucking) have _______________ value, some (such as stepping) form the basis for _____________________ that will develop later, and others (such as rooting and grasping) contribute to the infant's early____________________ ______________________________. (107)

3. Reflexes provide one way of assessing the health of a baby's ___________________ ______________. (107)

4. List the five states of arousal which newborn babies move in and out of throughout the day and night. (107-108)

A. ______________________________ B. ______________________________
C. ______________________________ D. ______________________________
E. ______________________________

5. True or False: These states of arousal rarely alternate during the infant's first month. (107)

6. During REM sleep, electrical brain wave activity is remarkably similar to that of ______________________________. During NREM sleep, the body is _______________________ and heart rate, breathing, and brain wave activity are __________________________________. (107-108)

7. REM sleep accounts for _____ percent of the newborn baby's sleep time in comparison to _____ percent in older children and adults. (108)

8. Why do young infants spend so much time in REM sleep? (108)

____
____
____

9. The powerful effect of the infants' cry on arousal and discomfort in adults is adaptive because it ensures that ____. (108)

10. The most effective technique for soothing a crying newborn when feeding and diaper changing do not work is ____. (109)

11. Like reflexes and sleep patterns, an infant's cry offers a clue to ________________ ______________________________ distress. The cries of babies who have experienced brain damage or prenatal and birth complications are often ___________________ and ___________________. This is one factor which contributes to the risk that parents will _________________________________. (109)

12. True or False: Sensitivity to touch is well developed at birth. (110)

13. Newborn infants are more sensitive to stimuli that are (hotter/colder) than body temperature. (110)

14. True or False: Newborn babies cannot feel pain. (110)

15. True or False: Newborn babies can distinguish among several tastes, and they prefer sweetness. (110)

16. True or False: Infants do not show odor preferences until they are several months old. (110)

17. Breast-fed babies recognize and prefer the odor of ___________________. Bottle-fed babies prefer the smell of ___________________________________ to their familiar formula. (110)

18. Newborn infants prefer _________________ sounds, such as noises and voices, to ___________________. (110)

19. True or False: Newborn babies can discriminate almost all speech sounds in human languages. (110-111)

20. List the special characteristics of adult speech that babies prefer. (111)

____

____

21. Vision is the (least/most) developed of the senses at birth. Briefly describe the newborn baby's visual acuity. (111)

____

____

22. True or False: Color vision is well developed at birth. (111)

23. The most widely used instrument for assessing the behavior of the infant during the newborn period is ____ __________, which evaluates ______________________, ______________________, ________________________________, and ____________________________. (111)

24. What have researchers learned from cross-cultural studies using the NBAS in the following cultures? (111)
Asian/Native American vs. Caucasian:______________________________

____

____

Zambia, Africa:______________________________

____

____

25. Rather than a single score, ______________________________ provide the best estimate of the baby's ability to recover from stress at birth. (111)

26. ______________________________ are clearly useful in helping the parent-infant relationship get off to a good start. (111-112)

Adjusting to the New Family Unit

1. List three adjustments families must make after the arrival of a new baby. (112)
A. ______________________________
B. ______________________________
C. ______________________________

ASK YOURSELF . . .

Amy, who is 2 months pregnant, wonders how the embryo is being fed and what parts of the body have formed. Amy imagines that very little development has yet taken place. How would you answer Amy's questions? Will she be surprised at your response? (see text pages 82-84)

____

____

____

____

How does brain development relate to fetal behavior during the third trimester? (see text pages 84-86)

____

____

____

____

Why is it difficult to determine the effects of some environmental agents, such as over-the-counter drugs and pollution, on the embryo and fetus? (see text pages 87-89)

____

____

Trixie has just learned that she is pregnant. Since she has always been healthy and feels good right now, she cannot understand why the doctor wants her to come in for checkups so often. Why is early and regular prenatal care important for Trixie? (see text pages 95-96)

____

____

____

Cite examples of prenatal environmental influences that appear to have little or no negative impact on health in infancy and childhood but that may contribute to serious health problems in adulthood. (see text pages 86-95)

What factors help the newborn baby withstand the trauma of labor and delivery? (see text page 98)

Use of any single medical intervention during childbirth increases the likelihood that others will also be used. Provide as many examples as you can to illustrate this idea. (see text pages 100-101)

How have history-graded influences (see Chapter 1, page 12) affected approaches to childbirth in Western nations? What effects have these changes had on the health and adjustment of mothers and newborn babies? (see text pages 99-101)

Sensitive care can help preterm infants recover, but unfortunately they are less likely to receive this kind of care than are full-term newborns. Explain why. (see text pages 102-103)

List all the factors discussed in this chapter that increase the chances that an infant will be born underweight. How many of these factors could be prevented by better health care for mothers and babies? (see text pages 102-103)

How do long-term outcomes repeated for preterm and low-birth-weight newborns fit with findings of the Kauai study? (see text pages 102-105)

Cite examples of how infant reflexes, state changes, and sensory capacities promote survival and provide a foundation for human motor, cognitive, and social capacities that will develop later. (see text pages 111-112)

Jackie, who had a difficult delivery, observes her 2-day-old daughter Kelly being given the NBAS. Kelly scores poorly on many items. Jackie wonders if this means that Kelly will not develop normally. How would you respond to Jackie's concern? (see text pages 107-108)

SUGGESTED READINGS

The American College of Obstetricians and Gynecologists. (1997). (2nd ed.). *Planning for pregnancy, birth, and beyond.* Washington, DC: The American College of Obstetricians and Gynecologists. Provides current information about preconception, prenatal, and postpartum health, including genetic disorders and birth defects, prenatal care, pregnancy complications, physical and emotional changes during pregnancy, labor and delivery, and the newborn baby.

Barr, R. G., Hopkins, B., & Green, J. A. (Eds.). (2000). *Crying as a sign, a symptom, and a signal: Clinical, emotional, and developmental aspects of infant and toddler crying.* New York: Cambridge University Press. Takes a multidisciplinary look at this important infant behavior, describes normative developmental patterns of crying, and discusses how they are manifested in various settings—emergency department, painful procedures, colic, temper tantrums, and nonverbal and mentally challenged infants.

Kotch, J. B. (1997). *Maternal and child health—programs, problems, and policy in public health.* Gaithersburg, MD: Aspen. Provides an overview of the health care needs of pregnant women, mothers, and children of all ages. Particular attention is focused on services, policies, and institutional forces that directly influence maternal and child health. Weaknesses and strengths of current federal and state policies and health reform issues are discussed.

Moore, K. L., & Persaud, T. V. N. (1998). *Before we are born* (5th ed.). Philadelphia: Saunders. A detailed presentation of prenatal development, emphasizing basic facts and concepts of normal and abnormal growth from a biological perspective.

Shepard, T. H. (1998). *Catalog of teratogenic agents* (5th ed.). Baltimore: Johns Hopkins University Press. This reference book documents the effects of a range of teratogens, including chemicals, drugs, physical factors, and viruses. Also described is research related to various teratogens and information on the species studied, dose, gestational age at time of administration, and type of congenital defects produced.

PUZZLE 3.1 TERM REVIEW

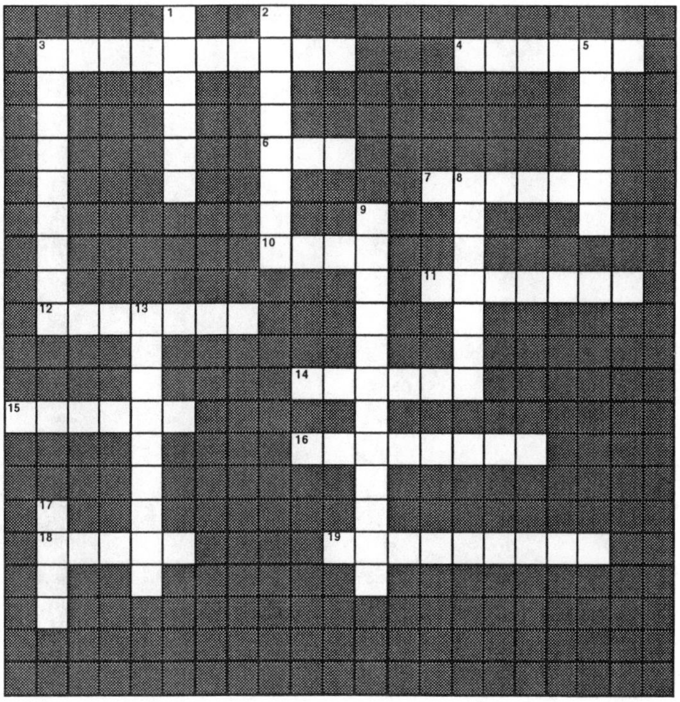

Across

3 Three month time periods of pregnancy
4 Inborn, automatic response to a stimulus
6 "Irregular" sleep state in which brain wave activity is similar to that of the waking state; eyes dart beneath the lids, heart rate, blood pressure, and breathing are uneven, and slight body movements occur (abbr.)
7 Visual _____: fineness of discrimination
10 "Regular" sleep state in which the body is quiet and heart rate, breathing, and brain wave activity are slow and regular (abbr.)
11 Born several weeks before due date
12 Childbirth approach designed to reduce pain and medical intervention and provide a rewarding experience
14 Membrane that serves as the protective outer layer that encloses the developing organism in amniotic fluid
15 Inadequate oxygen supply
16 Separates the mother's bloodstream from the fetus's, but permits exchange of nutrients and waste
18 _____ scale: rating used to assess newborn condition immediately after birth
19 Low birth weight is a major cause of infant _____.

Down

1 White, cheeselike substance that covers fetus and prevents chapping
2 Surgical delivery in which an incision is made in the mother's abdomen and the baby is lifted out of the uterus
3 Environmental agent that causes damage during the prenatal period
5 Prenatal organism from 2 to 8 weeks after conception, when foundations of all body structures and organs are laid down
8 Outer membrane that forms a protective covering and sends out villi from which the placenta emerges
9 Attachment of the blastocyst to the uterine lining 7 to 9 days after conception
13 Cord connecting the fetus to the placenta
17 Small-for-_____: below normal birth weight in relation to length of pregnancy

PUZZLE 3.2 TERM REVIEW

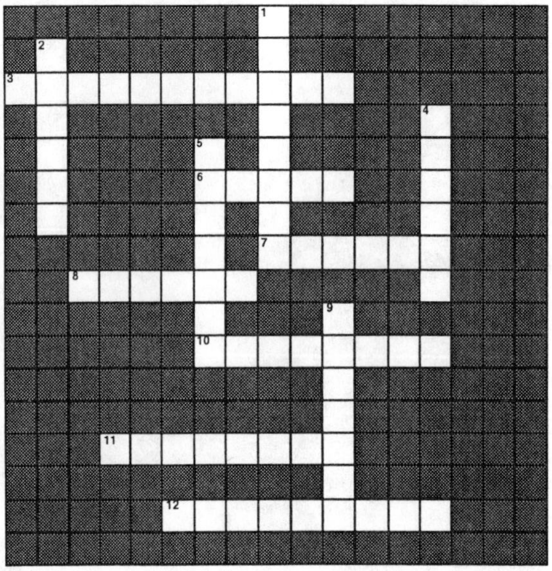

Across

3 _____ distress syndrome: a disorder of preterm infants in which the lungs are so immature that the air sacs collapse

6 Period of the _____ lasts from the ninth week to the end of pregnancy.

7 White, downy hair that helps the vernix stick to the skin

8 Position of the baby in the uterus in which the buttocks or feet would be delivered first

10 Fetal alcohol _____: mental retardation, slow growth, and facial abnormalities resulting from consumption of large amounts of alcohol during pregnancy

11 Fetal _____: electronic instruments that track the baby's heart rate during labor

12 Age of _____: age at which the fetus can first survive if born early, between 22 and 26 week

Down

1 _____ Behavioral Assessment Scale: test developed to assess the behavior of the infant during the newborn period

2 _____ tube: primitive spinal cord developing from the ectoderm, the top of swells to form the brain

4 Rh _____: protein that, when present in the fetus's blood but not in the mother's may cause the mother to build up antibodies which can destroy the fetus's red blood cells

5 Fetal alcohol _____: condition of children who display some but not all of the defects of fetal alcohol syndrome

9 States of _____: different degrees of sleep and wakefulness

SELF-TEST

1. By the fourth day after conception, 60 to 70 cells form a hollow, fluid-filled ball called a(n) (82)
   a. blastocyst.
   b. embryonic disc.
   c. zygote.
   d. chorion.

2. The organ that separates the mother's bloodstream from the embryo or fetal bloodstream, but permits exchange of nutrients and waste, is called the (82)
   a. amnion.
   b. chorion.
   c. placenta.
   d. umbilical cord.

3. Glial cells, which support and feed the brain's neurons (84)
   a. are all in place by the end of the second trimester.
   b. continue to increase rapidly throughout pregnancy and after birth.
   c. do not contribute to the development of new behavioral capacities.
   d. develop at the same rate as neurons.

4. Which of the following is true about the impact of teratogens? (86-87)
   a. They exert their most harmful effects during the period of the fetus.
   b. Their influence is unaffected by genetic makeup of mother and baby.
   c. Their effects on development are always direct.
   d. They tend to do most damage during the period of the embryo.

5. The repeated use of aspirin during pregnancy is related to (88-89)
   a. low birth weight, infant death, poorer motor development, and lower IQ scores in early childhood.
   b. irritability, vomiting, and infant jitteriness.
   c. no known ill effects.
   d. high birth weight.

6. What amount of alcohol is required to produce FAE? (91)
   a. extreme maternal alcoholism
   b. more than two drinks a day
   c. as little as one drink per week
   d. the precise amount has not been determined

7. Which of the following is NOT true of infants exposed to radiation? (91-92)
   a. They never appear normal at birth.
   b. They may have lower intelligence and/or emotional disorders.
   c. They may have underdeveloped brains.
   d. They may have physical deformities.

8. Pregnant women should not eat under-cooked meat or clean a cat's litter box due to the danger of (92-93)
   a. toxoplasmosis.
   b. rubella.
   c. AIDS.
   d. tuberculosis.

9. The effects of prenatal malnutrition (93-94)
   a. are minimal.
   b. cannot be reversed after birth.
   c. include damage to the CNS and other parts of the body.
   d. depend on maternal age.

10. Which of the following is NOT a barrier to early prenatal care in the United States (95)
    a. lack of health insurance for low-income families
    b. psychological stress
    c. demands of taking care of other young children
    d. inadequate health care services.

11. High levels of stress hormones produced by the baby during labor contribute to (98)
    a. the possibility of brain damage.
    b. difficulty breathing at the time of birth.
    c. the baby's ability to withstand oxygen deprivation.
    d. sleep.

12. The _____________ is used to assess the physical condition of the newborn at 1 and 5 minutes after birth. (98-99)
    a. Brazelton Neonatal Behavioral Assessment Scale
    b. Apgar Scale
    c. Bayley Scales of Infant Development
    d. Neonatal Reflexive Inventory

13. Research suggests that __________ may be an important factor in the success of natural childbirth methods. (100)
    a. labor and delivery medications
    b. the training of attending physicians
    c. social support
    d. correct use of breathing techniques

14. Home delivery (100)
    a. is just as safe as giving birth in a hospital for all women.
    b. is more dangerous than giving birth in a hospital for all women.
    c. is just as safe as giving birth in a hospital for healthy, well-assisted women.
    d. is more dangerous if the attendant is a trained nurse-midwife rather than a doctor.

15. Anoxia during labor and delivery may lead to (100-101)
    a. Down syndrome.
    b. cerebral palsy.
    c. small for dates.
    d. postterm births.

16. Fetal monitors (101)
    a. reduce the rate of infant brain damage in healthy pregnancies.
    b. are linked to an increased rate of cesarean deliveries.
    c. are unlikely to be used routinely due to their drawbacks.
    d. are not beneficial in high-risk situations.

17. Dawn was born after 38 weeks and weighed 3 1/2 pounds. Today, she would be classified as (102)
    a. small for date.
    b. preterm.
    c. premature.
    d. normal.

18. Which of the following statements is NOT true? (102-103)
    a. The appearance and behavior of preterm babies can affect the kind of care they receive.
    b. Research indicates that, compared to full term babies, preterm babies are less often held close, touched, and talked to.
    c. How well preterm babies develop depends to a great extent on the quality of the parent-infant relationship.
    d. Physical stimulation is usually harmful to fragile preterm infants.

19. Interventions that lead to improvements in the parenting and development of preterm infants include those which (103)
    a. teach parents about infant characteristics and caregiving skills.
    b. combine medical follow-up, parent training, and stimulating daycare.
    c. help infants sleep longer.
    d. do both a and b.

20. REM sleep (107-108)
    a. accounts for less time in preterm infants than in full term infants.
    b. is less frequent in newborns, in general, because they are already over-stimulated by their surroundings.
    c. is the stage of sleep in which there is little motor activity and breathing is slow and regular.
    d. seems to fulfill a need for central nervous system stimulation, especially in newborns.

21. The stepping reflex (106)
    a. is unrelated to walking.
    b. disappears sooner in babies who gain weight rapidly.
    c. does not occur under water.
    d. does not occur in babies younger than two months of age.

22. Observations of the sleep states of newborn babies can help identify: (108)
    a. dreams.
    b. hunger patterns.
    c. central nervous system abnormalities.
    d. need for stimulation.

23. Sensitivity to touch, especially around the mouth, is well developed (110)
    a. shortly after birth.
    b. at birth.
    c. by three months.
    d. by six months.

24. Because the visual system is not well developed, ___________, or fineness of discrimination, is limited in newborn babies. (111)
    a. depth perception
    b. visual acuity
    c. tracking ability
    d. linear perspective

25. Which of the following is true? (111-112)
    a. A single score on the NBAS is a good predictor of later development.
    b. The NBAS cannot determine the effects of child-rearing practices.
    c. The NBAS can predict future intellectual developments.
    e. The NBAS is useful only for assessment of children in the United States.

CHAPTER 4

PHYSICAL DEVELOPMENT IN INFANCY AND TODDLERHOOD

BRIEF CHAPTER SUMMARY

Body size increases dramatically during the first 2 years of life, following organized patterns of growth called cephalocaudal and proximodistal trends. Fat increases much more rapidly than does muscle during the first year. During the first 2 years, neurons form intricate connections, and their fibers myelinate, leading to a rapid increase in brain weight. Already, the two hemispheres of the cortex have begun to specialize, although the brain retains considerable plasticity during the first year of life. Researchers have identified key, sensitive periods of brain development when appropriate stimulation is key to acquiring skills.

A variety of factors affect early physical growth. While heredity contributes to height, weight, and rate of physical maturation, good nutrition is essential for rapidly growing babies and breast milk is especially well-suited to meet their needs. Malnutrition during the early years can lead to serious dietary diseases, which are associated with permanent stunting of physical growth and brain development. Finally, affection and stimulation are also vital for healthy physical growth.

Infants are marvelously equipped to learn immediately after birth. Classical and operant conditioning, habituation-dishabituation, and imitation are important mechanisms through which infants learn about their physical and social worlds.

The rapid motor development that occurs during the first 2 years follows the same organized sequences as does physical growth. The mastery of motor skills involves acquiring increasingly complex, dynamic systems of action. In this way, maturation and experience combine to influence the development of motor skills.

Perception changes remarkably over the first year of life. Hearing and vision undergo major advances during the first 2 years as infants organize stimuli into complex patterns, improve their perception of depth and objects, and combine information across sensory modalities. The Gibsons' differentiation theory helps us understand the course of perceptual development.

LEARNING OBJECTIVES

After reading this chapter, you should be able to:

4.1. Describe changes in body size, muscle-fat makeup, and proportions during the first 2 years, along with individual and group differences. (118-120)

4.2. Describe brain development during infancy and toddlerhood, at the level of individual brain cells and at the level of the cerebral cortex. (120-124)

4.3. Describe changes in the organization of sleep and wakefulness during the first two years, noting the contributions of brain maturation and the social environment. (124-126)

4.4. Describe evidence indicating that heredity contributes to body size and rate of physical growth, and discuss the nutritional needs of infants and toddlers. (126-127)

4.5. Describe two dietary diseases caused by malnutrition during infancy and toddlerhood, along with their consequences for physical growth and brain development. (128-129)

4.6. Discuss the origins and symptoms of nonorganic failure to thrive. (129)

4.7. Describe four infant learning capacities, the conditions under which they occur, and the unique value of each. (129-132)

4.8. Describe the sequence of motor development during the first 2 years. (133-134)

4.9. Discuss research indicating that motor skills are dynamic systems jointly influenced by factors internal and external to the child. (134)

4.10. Describe the development of voluntary reaching and grasping, citing evidence on the influence of early experience. (135-136)

4.11. Summarize the development of hearing and vision during infancy, giving special attention to depth and pattern perception. (136-141)

4.12. Describe evidence indicating that, from the start, babies are capable of intermodal perception. (142)

4.13. Explain differentiation theory of perceptual development. (142-143)

STUDY QUESTIONS

Body Growth

1. By the end of the first year, the infant is ____ percent longer than it was at birth, and by 2 years it is _____ percent longer. By 5 months, birth weight has ______________, at 1 year it has tripled, and at 2 years it has ______________. (118)

2. True or False: Researchers have found that infant and toddler growth takes place in spurts: as much as a half-inch in a 24-hour period. (118)

3. Baby fat helps the small infant ______________________________________. During the second year, toddlers ____________________, a trend that continues into middle childhood. ________________ tissue increases very slowly during infancy and does not peak until adolescence. (118)

4. True or False: Boys and girls are equal in ratio of fat to muscle during infancy. (119)

5. The best way of estimating a child's physical maturity is to use ___________ ________________________, a measure of bone development determined through X-rays. When the skeletal ages of infants and children are examined, _________________ children are slightly ahead of Caucasian children and (girls/boys) are considerably ahead of (girls/boys). (119)

6. Growth progressing from head to tail represents the ____________________ trend. Growth proceeding from the center of the body outward represents the ____ trend. (120)

Brain Development

1. The human brain has 100 to 200 billion ______________. Between them are tiny gaps, or ______________, across which chemical messages pass. (120)

2. True or False: During the prenatal period, more neurons than the brain will ever need are produced. (120)

3. As neurons form connections, ___________________ becomes important in their survival. (120)

4. During a process called _________________________, neurons seldom stimulated lose their synapses. (120)

5. About _________ of the brain's volume is made up of glial cells. Explain their function, using the term *myelinization* in your answer. (121)

____

____

6. True or False: The cerebral cortex is probably more sensitive to environmental influences than any other part of the brain. (121)

7. The order in which cortical regions develop corresponds to the order in which ____. (121)

8. Describe physical abnormalities associated with sudden infant death syndrome (SIDS) which may prevent these babies from learning how to respond when their survival is threatened. (123)

____

____

____

9. ___________________________ is the leading cause of infant mortality in the United States. (123)

10. List three environmental factors associated with SIDS. (123)
A. ________________________________ B. _________________________________
C. ________________________________

11. For most people the left hemisphere of the cortex is responsible for _____________________ and ______________________, while the right hemisphere is responsible for ______________________ and ______________________. (122)

12. Describe evidence for early plasticity in brain development. (122)

____

____

____

13. What is the adaptive function of brain lateralization? (122-123)

____

14. Give an example of the existence of sensitive periods in development of the cerebral cortex, as demonstrated in studies of animals exposed to extreme forms of sensory deprivation. (123-124)

____

____

____

15. Ethically, researchers cannot expose children to extreme sensory deprivation. However, less direct evidence closely parallels the animal evidence. List two examples of evidence in humans of the existence of sensitive periods in development of the cerebral cortex. (124)

A.____

____

B.____

____

16. List two types of environments that fail to capitalize on children's brain potential. (124)

A.____

B.____

17. Describe the greatest change in sleep and wakefulness between birth and 2 years. (124)

____

____

18. In contrast to the nighttime separation which occurs in many American families, many cultures around the world practice parent-infant ____________ . (125)

19. True or False: Infant sleeping arrangements around the world are primarily determined by availability of household space. (125)

20. True or False: Bedtime struggles are a universal phenomenon. (125)

Influences on Early Physical Growth

1. Describe the phenomenon of *catch-up growth.* What other evidence is there for the heritability of growth tendencies? (126)

Catch-up growth: ____

____

Other evidence:____

____

2. List four major nutritional and health advantages of breast milk. (126-127)

A. ____

B. ____

C. ____

D. ____

3. Explain why breast-fed babies become hungry about twice as often as bottle-fed infants. (127)

____

4. True or False: Most chubby infants become obese children and adults. (127-128)

5. Cite two ways in which parents can prevent infants and toddlers from becoming overweight at later ages. (128)
A. ____
B. ____

6. By middle childhood, children who survive extreme forms of malnutrition score low on __________________________, show poor __________________________ __________________, and have difficulty ____________________________. (128)

7. Describe the cause and symptoms of *nonorganic failure to thrive,* and its consequences if not treated early. (129)
Cause: ____
Symptoms: ____
____
Consequences:____
____

Learning Capacities

1. *Learning* refers to changes in __________________________ as the result of __________________________. (129)

2. Briefly explain why classical conditioning is of great value to infants. (130)

____

3. In classical conditioning: (130)
a) an ____________________________ stimulus must consistently produce a reflexive, or ____________________________ response.
b) Then a __________________ stimulus that does not lead to the reflex is presented with the ____________________________ stimulus.
c) Finally, the neutral stimulus, or ____________________ stimulus produces the reflexive response, now referred to as the ____________________ response.

4. Under what conditions can young infants most easily be conditioned? What types of responses are difficult to condition? (130)
Easy: ____
Difficult: ____

5. In *operant conditioning,* infants __________ on the environment and then are influenced by the stimuli which follow. What effect do the following types of stimuli have on the probability that the preceding behavior will be repeated? (131)
Reinforcers: ____
Punishers:____

6. Describe the role that operant conditioning plays in maintaining pleasurable interactions between parent and baby. (131)

____

____

7. Describe *habituation* and *dishabituation*. (131)
Habituation:____

____

Dishabituation: ____

____

8. Habituation and dishabituation allow us to focus our attention on those aspects of the environment that ______________________. (131)

9. List some responses that newborns can imitate. (132)

____

____

10. Explain how imitation provides the young baby with a powerful means of learning. (132)

____

____

____

Motor Development

1. Distinguish between gross and fine motor development. (133)
Gross: ____

____

Fine: ____

____

2. True or False: Although the *sequence* of motor development is fairly uniform, there are large individual differences in *rate* of motor progress. (134)

3. Motor development is a matter of acquiring increasingly complex __________ ______________________. (134)

4. Why can motor development be mapped out by heredity only at a very general level? (134)

____

5. Briefly describe Dennis' s findings on the movement opportunities of Iranian institutionalized babies, which support the role of experience in early motor development. (134)

____

____

____

6. Briefly explain what factors account for the advanced motor development of Kipsigis and West Indian infants. (134)

____
____
____

7. Of all motor skills, ______________________________ is believed to play the greatest role in infant cognitive development. (135)

8. Describe four main steps in the development of reaching which are mastered around the following ages: (135)
Newborn: ____
3 months: ____
5 months: ____
7 months: ____
9 months: ____

9. By the end of the first year, infants use thumb and index finger in a well-coordinated _____________ grasp. (135)

10. Describe well-known evidence indicating that voluntary reaching is influenced by early experience. (136)

____
____
____

Perceptual Development

1. What is the greatest change that takes place in hearing over the first year of life? (136)

____

2. True or False: Infants generally cannot identify the precise location of a sound before one year of age. (136)

3. In preparation for language acquisition, infants are able to "screen out" ____ by 6 months. (136)

4. In the second half of the first year, infants focus on _____________________ ___________ that are critical to understanding the meaning of what they hear. (136)

5. List four changes in infants' ability to see clearly and explore the visual field that result from rapid maturation of the eye and visual centers in the brain. (137)
A. ____
B. ____
C. ____
D. ____

6. Depth perception is the ability to judge ______________________________ ____.
It is important for guiding _____________ activity. (137)

7. Based on the visual cliff studies, researchers concluded that around the time that infants crawl, most distinguish ____________________ and ____________________ surfaces and avoid ______________________________. (137)

8. Name and describe three cues for depth. Indicate the approximate age at which infants become sensitive to each. (137-138)
A.____
____
B.____
____
C.____
____

9. True or False: Greater crawling experience is related to avoidance of dangerous heights as well as the ability to find hidden objects. (138)

10. Provide an example from adult experience which helps explain why crawling plays such an important role in the infant's knowledge and understanding of the three-dimensional world. (138-139)
____
____

11. List the three aspects of development that infants with severe visual impairments showed delays. (138)
A.______________________________ B.______________________________
C.______________________________

12. Minimal or absent vision impacts infants' ________________________ exploration and spatial ______________________________. (138)

13. True or False: Infants with visual impairments usually receive more adult attentions, play, and other stimulation than sighted infants. (138)

14. Once ____________________ emerges and the child can rely on it for learning, some children with limited or no vision show impressive rebounds. (139)

15. With age, infants spend more time looking at patterns with fine details. Name and describe the principle that accounts for this trend. (140)
____
____

16. At about _____ months of age, infants start to thoroughly explore a pattern's internal features. By the age of __________ months, infants are so good at detecting pattern organization that they even perceive subjective boundaries that are not really present. (140)

17. The baby's tendency to search for structure in a patterned stimulus applies to face perception. By _____ months, infants make fine-grained distinctions among the features of different faces. Between _____ to _____ months, infants start to perceive facial expressions as organized wholes. (141)

18. From the start, infants perceive the world in an ____________________ fashion. Explain what this means. (142)

____

____

19. How many exposures do infants typically require to pick up an intermodal association? __________ (142)

Understanding Perceptual Development

1. According to the Gibsons' differentiation theory, the search for ________________ ________________ of the environment, or those that remain ____________________, provides an explanation of infants' perceptual development. (142)

2. One way of understanding perceptual development is to think of it as a built-in tendency to look for ________________________________, a capacity that becomes more ________________ with age. (143)

3. True or False: According to the Gibsons, opportunities to act on the environment play a major role in development. (143)

ASK YOURSELF . . .

Felicia commented that at 2 months, April's daily schedule seemed more predictable and she was much more alert. What aspects of brain development might be responsible for this change? (see text pages 120-124)

____

____

We are used to thinking of brain development as following a strict, genetically determined course, but research shows that this is not so. List examples that reveal the effects of experience on brain growth and lateralization of the cerebral cortex. (see text pages 122-123)

____

____

____

____

Why is lateralization of the cerebral cortex adaptive for lifelong learning? (123-124)

____

____

____

Explain why breast-feeding can have lifelong consequences for the development of babies born in poverty-stricken regions of the world. (see text page 127)

____

____

____

Ten-month-old Shaun is below average in height and painfully thin. He has one of two serious growth disorders. Name them, and indicate what clues you would look for to tell which one Shaun has. (see text pages 128-129)

A.____

B.____

Nine-month-old Byron has a toy with large, colored push buttons on it. Each time he pushes a button, he hears a nursery tune. Which learning capacity is the manufacturer of this toy taking advantage of? What can Byron's play with the toy reveal about his perception—specifically, his ability to distinguish sound patterns? (see text pages 129-132, 136-137)

learning capacity: ________________________________

perception:____________________________________

Earlier in this chapter, we indicated that infants with nonorganic failure to thrive are unlikely to smile at a friendly adult. Also, they keep track of nearby adults in an anxious and fearful way. Explain these reactions using the learning capacities discussed in the preceding sections. (see text pages 129-132)

Review the new approach to motor development (see text page 134), which regards motor skills as complex systems of action. Is motor development multidimensional? Explain. (Return to Chapter 1, page 9, if you need to review this assumption of the lifespan perspective.)

Rosanne read in a magazine that infant motor development could be accelerated through exercise and visual stimulation. She hung mobiles and pictures all over her newborn baby's crib, and she massages and manipulates his body daily. Is Rosanne doing the right thing? Why or why not? (see text page 136)

Five-month-old Tyrone sat in his infant seat, passing a teething biscuit from hand to hand, moving it up close to his face and far away, and finally dropping it overboard onto the floor below. What aspect of visual development is Tyrone probably learning about? Explain your answer. (see text pages 137-139)

____

____

____

Diane put up bright wallpaper with detailed pictures of animals in Jana's room before she was born. During the first 2 months of life, Jana hardly noticed the wallpaper. Then, around 2 months of life, she showed keen interest. What new visual abilities probably account for this change? (see text page 140)

____

____

____

SUGGESTED READINGS

Adoph, K. E. (Ed.). (1997). *Learning in the development of infant locomotion.* Monographs of the Society for Research in Child Development, 62 (3, Serial No. 251). Chicago: University of Chicago Press. Examines infants' adaptive responses in a novel task—going up and down slopes—and reveals a process of differentiation and selection caused by changes in everyday experience, body dimensions, and locomotor proficiency.

Bremner, J. G., Slater, A., & Butterworth, G. (Eds.). (1997). *Infant development: Recent advances.* Washington, DC: Psychology Press. An edited volume that provides most current research and theoretical knowledge concerning perceptual, cognitive, and social development in infancy. Also focuses on the links between these developmental domains. Compares and contrasts different perspectives in a highly readable, accessible manner.

Dawson, G., & Fischer, K. W. (1999). *Human behavior and the developing brain.* New York: Guilford Press. Describes research and theory on developmental changes in the brain. Discusses neural correlates of developmental processes pertaining to memory, emotional expression, spatial representation, and language.

Mehler, J., & Dupoux, E. (1994). *What infants know: The new cognitive science of early development.* Cambridge, NJ: Blackwell Publishers. Discusses the extent to which infants are "pre-wired" for various perceptual and cognitive capacities. Debates what it means to be "born knowing" and proposes a new conceptualization of human development.

Slater, A. (1998). *Perceptual development: Visual, auditory, and speech perception in infancy.* Philadephia, PA: Psychology Press. Provides an overview of perceptual development of visual, auditory, and speech perception in infancy.

PUZZLE 4.1 TERM REVIEW

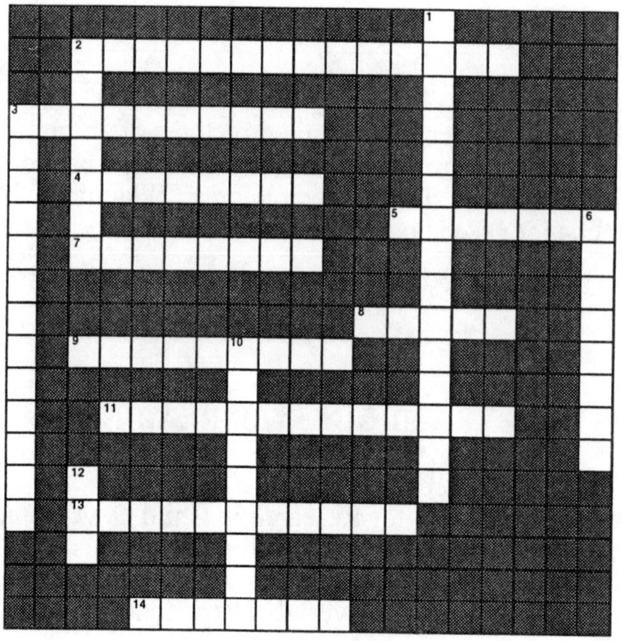

Across

2 A change in the environment causes responsiveness to a stimulus to return to a high level.
3 Decreases likelihood of repeating a response
4 A disease caused by a diet low in all essential nutrients
5 Cells that store and transmit information
7 _____ cortex: largest, most complex brain structure
8 Brain cells responsible for myelinization
9 Form of conditioning involving association of a neutral stimulus with one that leads to a reflexive response
10 Process of coating neural fibers with fatty sheath
13 _____ stimulus: a neutral stimulus that, when paired with an unconditioned stimulus, leads to a conditioned response
14 Synaptic _____: process in which neurons that are seldom stimulated lose their synapses

Down

1 According to _____ theory, perceptual development is a matter of detecting invariant features in a constantly changing perceptual world.
2 _____ systems of action involve combinations of previously acquired abilities
3 Pattern of growth proceeding from the center of the body outward
6 Gap between neurons
10 Learning by copying
12 A stimulus that leads to a reflexive response (abbr.)

PUZZLE 4.2 TERM REVIEW

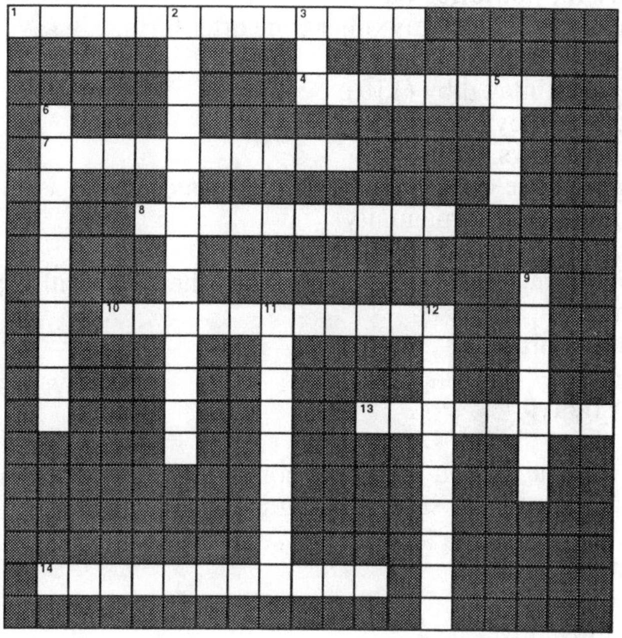

Across

1 Growth that proceeds from head to tail
4 Conditioned _____: produced by the neutral stimulus after the neutral stimulus has been paired with a unconditioned stimulus
7 _____ failure to thrive: growth disorder resulting from lack of parental love
8 Ability of other parts of the brain to take over the functions of damaged regions
10 A disease caused by a diet low in protein
13 _____ sensitivity accounts for early visual pattern preferences
14 Reduction in the strength of a response due to repetitive stimulation

Down

2 Specialization of brain hemispheres
3 A reflexive response produced by an unconditioned stimulus (abbr.)
5 Abbreviation for death of a seemingly healthy baby without apparent cause
6 Perception combining information from more than one sensory system
9 A form of conditioning in which spontaneous behaviors are followed by stimuli that change the probability that the behavior will be repeated
11 Features that remain stable
12 A stimulus that increases the occurrences of a response

SELF-TEST

1. During the first 2 years after birth, physical growth is (118)
   a. more rapid than it was during the prenatal period.
   b. slow and steady.
   c. rapid only for bottle-fed babies.
   d. faster than it will be at any other time after birth.

2. Skeletal age is estimated by (119)
   a. dividing height by age in years.
   b. examining X-rays.
   c. determining bone density through ultrasound.
   d. examining the joints manually.

3. Once neurons form connections in the brain, their survival depends on (120)
   a. stimulation.
   b. the cerebral cortex.
   c. myelinization.
   d. brain plasticity.

4. The structure of the human brain that contains the greatest number of neurons and synapses and is the last to stop growing is the (121)
   a. brain stem.
   b. cerebellum.
   c. cerebral cortex.
   d. corpus callosum.

5. What factor appears to be related to an increased risk of SIDS? (123)
   a. sleeping on the back
   b. a mother who smokes
   c. living in crowded conditions
   d. insufficient body warmth

6. Brain plasticity has to do with (122-123)
   a. the continued development of neurons.
   b. the ability of other parts of the brain to take over functions of a damaged region.
   c. the amount of stimulation the brain can process.
   d. premature closing of the fontanels.

7. Changes in sleep patterns over the first 2 years can best be explained by (124)
   a. brain development.
   b. the social environment.
   c. inhibition of the startle response.
   d. the social environment and brain maturation.

8. Which of the following is most consistent with the term *catch-up growth*? (126)
   a. A smaller identical twin becomes more like his/her larger twin over time.
   b. Girls tend to be advanced in growth during childhood, but boys outgrow girls during adolescence.
   c. Preterm babies gain more weight than full-term babies.
   d. Younger siblings are likely to outgrow older siblings.

9. Which of the following is NOT a nutritional or health advantage of breast milk? (126-127)
   a. high protein, low fat content
   b. digestibility
   c. protection against disease
   d. nutritional completeness

10. Longitudinal studies of the development of children who suffer prolonged malnourishment suggest that, given dietary intervention in preschool, (127))
    a. deficits are not long lasting.
    b. they may catch up in brain growth, but not in height.
    c. they may experience catch-up growth, but not in head size.
    d. cognitive deficits disappear.

11. Nonorganic failure to thrive (129)
    a. cannot be treated.
    b. is caused by lack of affection and stimulation.
    c. results from a failure to offer food to the infant.
    d. is a serious viral disease.

12. Which of the following is true? (129-130)
    a. Babies can be classically conditioned with any pairing of stimuli.
    b. Fear is one of the easiest responses to condition in young infants.
    c. Fear is difficult to condition in infants between 8 and 12 months of age.
    d. Classical conditioning is important because it allows infants to predict and make sense out of their environment.

13. Which of the following is true? (130-131)
    a. Newborns cannot learn through operant conditioning because they have no voluntary responses.
    b. Food is the only possible reinforcer for the young infant.
    c. It is easy to classically condition almost any stimulus pairing in young babies.
    d. Successful operant conditioning is limited to sucking and head turning responses in newborn infants.

14. After a baby's attention to a new stimulus declines, a second novel stimulus causes responsiveness to return to a high level. The infant is then demonstrating (131)
    a. classical conditioning.
    b. habituation.
    c. consequential learning.
    d. dishabituation.

15. The cephalocaudal trend is apparent in which of the following? (134)
    a. Motor control of the arms precedes control of the hands and fingers.
    b. Motor control of the legs precedes control of the arms.
    c. Motor control of the head precedes control of the arms.
    d. Motor control of the trunk precedes control of the head.

16. Kicking, rocking on all fours, and reaching are gradually put together into crawling. This example illustrates that motor development is a matter of acquiring increasingly complex (134)
    a. fine motor development.
    b. gross motor development.
    c. dynamic systems of action.
    d. coordinated action.

17. Newborns' uncoordinated reaching movements are called (135)
    a. prereaching.
    b. proprioceptive reactions.
    c. arm reflexes.
    d. ulnar grasps.

18. Over the first year of life, the greatest change in hearing that takes place is the ability to (136)
    a. hear low range tones.
    b. organize sounds into complex patterns.
    c. hear whispers.
    d. tolerate loud noises.

19. Visual acuity reaches a near-adult level at (137)
    a. 3 months of age.
    b. 6 months of age.
    c. 11 months of age.
    d. 18 months of age.

20. Infant depth perception, as assessed by refusal to cross the deep side of the visual cliff, is promoted by (138-139)
    a. a regimen of physical exercise.
    b. independent movement.
    c. opportunities to look at photographs and picture books.
    d. holding babies over drop-offs.

21. Severe visual impairment in infants impact all of the following, except (138-139)
    a. attainment of gross and fine motor skills.
    b. caregiver-infant relationships.
    c. cognitive development.
    d. language acquisition.

22. Changing visual pattern preferences during infancy are thought to be due to (140)
    a. improvements in contrast sensitivity.
    b. enlargement of the fovea.
    c. experience with complex patterns.
    d. reinforcements for attending to facial features.

23. At about what age do infants begin to thoroughly explore a pattern's internal features? (140-141)
    a. shortly after birth
    b. 2 months
    c. 6 months
    d. 1 year

24. From the start, infants expect sight, sound, and touch to go together, a capacity called (142)
    a. intermodal perception.
    b. system of action.
    c. differentiation organization.
    d. contrast sensitivity.

25. According to Eleanor and James Gibson, perceptual development is (142-143)
    a. a matter of detecting stable features in a constantly changing environment.
    b. a matter of stimulus-response learning.
    c. totally controlled by neural maturation.
    d. totally controlled by visual maturation.

CHAPTER 5

COGNITIVE DEVELOPMENT IN INFANCY AND TODDLERHOOD

BRIEF CHAPTER SUMMARY

According to Piaget, from earliest infancy children actively build schemes as they manipulate and explore their world. By acting on the world during the sensorimotor stage, infants make strides in intentional behavior and understanding of object permanence. In the last of six substages, toddlers transfer their action-based schemes to a mental level, and representation appears. Recent research, especially in the areas of physical reasoning and problem-solving, indicates that a variety of sensorimotor capacities emerge earlier than Piaget believed, raising questions about the accuracy of his account of sensorimotor development.

The information processing approach focuses on the development of mental strategies for storing and interpreting information. With age, infants attend to more aspects of their environment and remember information over longer periods of time. Findings on infant categorization suggest that babies structure experience in adultlike ways. Vygotsky's sociocultural theory stresses that cognitive development is socially mediated as adults help infants and toddlers master challenging tasks.

A variety of infant intelligence tests have been devised to measure individual differences in early mental development. Although most predict later performance poorly, those that emphasize recognition memory and object permanence show better predictability. Home and child care environments as well as early interventions for at-risk youngsters exert powerful influences on intellectual progress.

The rapid language development that occurs during the first 2 years is viewed very differently by behaviorist and nativist theories. Combining them, the interactionist view suggests that both innate abilities and environmental influences combine to produce children's language achievements. Infants prepare for language in many ways during the first year. Then first words appear around 12 months, 2-word utterances between 18 months and 2 years. However, substantial individual differences exist in rate and style of early language progress. Adults support youngsters' efforts to become competent speakers by using child-directed speech, a simplified form of parental language.

LEARNING OBJECTIVES

After reading this chapter, you should be able to:

5.1. Explain Piaget's view of what changes with development and how cognitive change takes place. (148-149)

5.2. Name Piaget's six sensorimotor substages, and describe the major cognitive achievements in each. (149-152)

5.3. Discuss recent research on sensorimotor development and its implications for the accuracy of Piaget's sensorimotor stage. (152-155)

5.4. Describe the structure of the information-processing system, the development of attention, memory, and categorization during infancy and toddlerhood, and the contributions and limitations of the information-processing approach to our understanding of early cognitive development. (156-160)

5.5. Explain how Vygotsky's concept of the zone of proximal development expands our understanding of early cognitive development. (160-161)

5.6. Describe the mental testing approach, the meaning of intelligence test scores, and the extent to which infant tests predict later performance. (162-163)

5.7. Discuss environmental influences on early mental development, including home, day care, and early intervention for at-risk infants and toddlers. (163-166)

5.8. Describe three major theories of language development, indicating the emphasis each places on biological and environmental influences. (166-168)

5.9. Describe how infants prepare for language, and explain how adults support their emerging capacities. (168-169)

5.10. Describe toddlers' first words and two-word combinations, and explain why language comprehension develops ahead of production. (169-170)

5.11. Describe individual differences in early language development and factors that influence these differences. (170)

5.12. Explain how child-directed speech, conversation, and reading to young children support early language development. (170-173)

STUDY QUESTIONS

Piaget's Cognitive-Developmental Theory

1. Piaget believed that ______________________________ change with age. He referred to specific structures as ________________. At first, these are ______________________________________. (148)

2. According to Piaget, one way schemes change is through ______________, a process that involves building schemes through direct interaction with the environment. (148)

3. Match the following terms with the appropriate descriptions and examples. (148-149)

_____ Creating new schemes or adjusting old ones to produce a better fit with the environment.
_____ Dropping various objects all in the same way.
_____ Using current schemes to interpret the external world.
_____ Modifying the way objects are dropped depending on their unique properties.

1. Assimilation
2. Accommodation

4. When children are not changing very much, they ________________________ more than they ________________________. They are in a state of ________________________. When children realize that new information does not fit current schemes, children are in a state of ________________________, or cognitive discomfort. (148-149)

5. Cognitive change can also occur internally, through a process of rearranging schemes called ________________________. (149)

6. Name and describe the special means that infants use to adapt their earliest schemes. (149)

____

____

____

7. Match each of the following sensorimotor substages with its appropriate description: (149-152)

____ Reflexes are the main means for responding to the world.
____ Infants try to repeat interesting effects in the surrounding world.
____ Exploration of the properties of objects by acting on them in novel ways
____ Infants' schemes are primarily oriented toward their own bodies.
____ Babies develop intentional, or goal-directed, behavior and object permanence.
____ Toddlers become capable of internal representation.

1. Reflexive Schemes
2. Primary Circular Reactions
3. Secondary Circular Reactions
4. Coordination of Secondary Circular Reactions
5. Tertiary Circular Reactions
6. Mental Combinations

8. During Substage ____, infants can successfully search for objects only in the first place in which they are hidden. They make the ________________________ error. (151)

9. Byron dropped objects down the basement steps, trying a variety of actions in a deliberately exploratory approach. Byron is demonstrating a ___________ circular reaction. (151)

10. List three signs that toddlers in Piaget's Substage 6 are capable of mental representation. (152)

A. ____

B. ____

C. ____

11. True or False: Piaget may have underestimated infant capacities. (152)

12. Explain why young infants (who appear to grasp the notion of object permanence) do not try to search for hidden objects. (152-153)

____

____

13. The ability to integrate object-location knowledge with action helps babies avoid the ________________ error. This ability may depend on rapid development of the frontal lobes of the cerebral cortex at the end of the ___________ year. (153)

14. Habituation and dishabituation research has revealed that very young infants are aware of object properties and the rules governing their behavior. List three basic regularities of the physical world that young infants grasp. (153)
A. ______________________________
B. ______________________________
C. ______________________________

15. Recent evidence indicates that the capacity for deferred imitation emerges by ____________________, not at the end of the sensorimotor stage, as Piaget believed. (153-154)

16. True or False: At the end of the second year, toddlers can imitate actions an adult tries to produce, even if these are not fully recognized. (154)

17. By 10 to 12 months, infants can solve problems by ____________________. (154)

18. Some schemes, such as ____________________ and ____________________ may be prewired into the brain from the start, while others may be constructed through ____________________. (155)

19. True or False: Research supports Piaget's view that infants' skills change together and abruptly as each new substage is attained. (155)

Information Processing

1. Describe the three basic parts of the information-processing system, including ways in which mental processes can facilitate the storage and retrieval of information at each level. (156-157)
Sensory Register: ______________________________
____
Working Memory: ______________________________
____
Long-term Memory: ______________________________
____

2. True or False: Information-processing researchers contend that the basic structure of the mental system is similar throughout life, but that its capacity increases. (157)

3. List two ways in which attention improves between one and five months of age. (157)
A. ______________________________
B. ______________________________

4. During the first year, infants attend to ____________________ events. In toddlerhood, children become more capable of ____________________ behavior. (157)

5. Like older babies and adults, babies seem to remember best when experiences take place in ____________________ contexts and when they ____________________ ____________________. (157)

6. Distinguish between *recognition* and *recall.* At what age are infants capable of recall? (157-158)
Recognition: ____
____
Recall: ____
____
Age at which infants can recall: ______________________________________

7. List two factors necessary for the development of autobiographical memory. (196)
A. ____
B. ____

8. The earliest categories are ________________, or based on similar overall appearance or prominent object part. By the end of the second year, categories are becoming ________________, or based on common function and behavior. (158)

9. By ________________ of age, infants structure objects and emotions into a wide array of categories. (158)

10. In the second year, children become _____________ categorizers during their __________. (158)

11. In what way does information-processing research challenge Piaget's view of early cognitive development? (158)

____
____
____

12. Cite the main drawback of the information-processing approach to cognitive development. (158-159)

____
____
____

The Social Context of Early Cognitive Development

1. According to Vygotsky's sociocultural theory, how do children come to master activities and think in culturally meaningful ways? (160)

____
____

2. Explain Vygotsky's concept of the *zone of proximal development* and how adults can foster development within it. (160)

____
____
____
____

3. Piaget concluded that toddlers discover make-believe play ______________, but Vygotsky believed that society provides children with opportunities to practice ______________________________ in play, first learned under the guidance of experts. (161)

4. True or False: When mothers take part, middle-class American toddlers' make-believe becomes more frequent and moves toward a more advanced level. (161)

Individual Differences in Early Mental Development

1. While the cognitive theories discussed so far try to explain the ___________ of development, mental tests are designed to measure cognitive ____________ and predict __________________________. (162)

2. Most infant intelligence tests consist of __________________ and __________ responses along with some tasks that tap early _____________________ and ____________________________. Name one commonly used infant test. (162)

____

3. When an intelligence test is constructed, it is given to a large sample of individuals whose performances form a _________________________________ and to whom future test-takers will be compared. (162)

4. A child who obtains an IQ of 100 scores higher than ______ percent of his or her agemates. A child who obtains a score of 130 scores higher than ________ percent of his or her agemates. (162-163)

5. Because of concerns about the predictability of infant tests, scores are labeled ___________________________________ rather than IQs. (163)

6. True or False: Because they show better long-term predictability for extremely low-scoring babies, infant tests are often used for screening of potential developmental problems. (163)

7. Infant tests based on Piaget's concept of object permanence are more predictive of future intelligence than are traditional tests because they tap the basic intellectual process of ____________________________. (163)

8. High HOME scores are associated with _________________________________, while low HOME scores predict ______________. (163)

9. The extent to which parents talk to infants and toddlers contributes to early language process, which in turn, predicts ______________________________________ in elementary school. (163)

10. What impact is intrusive parenting that is not sensitive to infants' and toddlers' ongoing actions likely to have on development? (164)

____

____

11. Today, over _____ of American mothers with children under age 2 are employed. (164)

12. Describe the overall quality of child care for infants and toddlers in the United States. (164)

____

____

____

13. Children who are most likely to be exposed to inadequate child care come from ____. (165)

14. List two main types of early intervention programs for infants and toddlers. (165)
A. ______________________________ B. ______________________________

15. Describe the child care treatment received by children in the Carolina Abecedarian Project. Did this treatment have an enduring impact on intelligence and achievement scores? (165-166)

____

____

____

____

Language Development

1. True or False: Conditioning and imitation are best viewed as supporting early language learning rather than fully explaining it. (167)

2. According to Chomsky, all children are born with a ____________________ ______________________________ that permits them, as soon as they have learned enough words, to speak in a rule-oriented fashion. (167)

3. List three challenges to Chomsky's theory. (167)
A. ____
____
B. ____
____
C. ____
____

4. According to the interactionist position, ______________________________, ______________________________, and ______________________________ combine to assist children in acquiring a language system. (167)

5. What evidence indicates that maturation cannot fully account for the development of babbling? (168)

____

____

6. Briefly explain how cooing and babbling contribute to preparation for language. (168)

____

____

7. Infants who experience joint attention with their caregivers (are/are not) likely to show faster language progress. (168)

8. Games such as pat-a-cake and peekaboo help babies practice the __________ ____________ pattern of conversation. (168)

9. At the end of the first year, as infants become capable of intentional behavior, __________________________ appear. Soon words are uttered along with them, and spoken language is underway. (169)

10. List the three subjects to which toddlers' first words typically refer. (168)
A. ______________________________ B. ______________________________
C. ______________________________

11. Provide an example of underextension and one of overextension. (169)
Underextension: ______________________________
Overextension: ______________________________

12. Toddlers' overextensions reflect their sensitivity to ____________________. (169)

13. True or False: At all ages, comprehension of language develops ahead of production. (170)

14. Toddlers' two-word utterances have been called telegraphic speech because they ______________________________. (170)

15. True or False: Early language development of boys and girls proceeds at about the same rate. (170)

16. Distinguish between referential and expressive styles of early language learning. Indicate which style is associated with faster vocabulary development. (170)
Referential: ______________________________
____
Expressive: ______________________________
____

17. Cite factors that influence the development of referential and expressive styles. (170)
____
____

18. Describe three ways caregivers can consciously support early language learning. (170-171)
A. ______________________________
B. ______________________________
C. ______________________________

19. Describe the characteristics of child-directed speech (CDS), a language style which supports early language development, and how it promotes language development. (171-172)
A.______________________________
____
____
B.______________________________
____

20. Explain why deaf children with hearing parents are often delayed in language progress and complex make-believe play, achieve poorly in school, and are deficient in social skills, whereas deaf children with hearing parents escape these difficulties. (172)
____
____
____
____

21. Why is amount of conversational give-and-take between parent and child an excellent predictor of early language development? (172-173)

____

____

ASK YOURSELF . . .

At 14 months, Tony pushed his toy bunny through the slats of his crib onto a nearby table. Using his "pulling scheme," he tried to retrieve it, but it would not fit back through the slats. Next Tony tried jerking, turning, and throwing the bunny. What kind of circular reaction is Tony demonstrating? How would Tony have approached this problem a few months earlier? (see text pages 150-151)

____

____

____

____

Mimi banged her rattle again and again on the tray of her high chair. Then she dropped the rattle, which fell out of sight on her lap, but Mimi did not try to retrieve it. Which sensorimotor substage is Mimi in? Why do you think so? (see text pages 150-151)

____

____

____

____

Three-year-old Katie solved an "animal" puzzle by matching the shape of each animal to the shape of the empty space on the puzzle board. Next, she tried a "vehicle" puzzle. "I know! Just match the shapes," she exclaimed and quickly solved the new puzzle. Does Katie's problem-solving strategy have roots in infancy? Explain. (see text pages 154-155)

____

____

____

____

Rachel played with toys in a more intentional, goal-directed way as a toddler than as an infant. What impact is Rachel's more advanced toy play likely to have on the development of attention? (see text page 157)

____

____

At age 18 months, Byron's father stood behind him, helping him throw a large rubber ball into a box. When Byron showed he could throw the ball, his father stepped back and let him try on his own. Using Vygotsky's ideas, explain how Byron's father is supporting his cognitive development. (see text pages 160-161)

____

____

____

____

Which theory of cognitive development can best explain how culturally relevant skills are passed from one generation to the next? Explain. (see text pages 160-161)

Fifteen-month-old Joey's DQ is 100. His mother wants to know what this means and what she should do at home to support his mental development. How would you respond to her questions? (see text pages 162-164)

Using what you learned about brain growth in Chapter 4, explain why intensive intervention for poverty-stricken children starting in the first 2 years has a greater long-term impact on IQ than intervention beginning at a later age. (see text pages 120-124)

Erin's first words included *see, give,* and *thank you,* and her vocabulary grew slowly during the second year. What style of early language learning did she display, and what factors might explain it? (see text pages 170)

Eighteen-month-old Toby lives in a poverty-stricken, overcrowded household. His parents seldom have time to converse with or to read to him. Summarize evidence indicating that Toby's limited early language acquisition may result in lasting deficits in development. (see text pages 170-173)

SUGGESTED READINGS

Bruer, J.T. (1999). *The myth of the first three years.* New York: Free Press. Challenges the widely accepted belief that the first 3 years of life determine whether or not a child will develop into a successful, thinking person. Drawing on research on brain development, shows that learning and cognitive development occur not just in infancy and toddlerhood but also throughout childhood and adulthood. Stresses the dangers of a view that overemphasizes early learning to the detriment of long-term parental and educational responsibilities.

Gopnik, A., Meltzoff, A.N., & Kuhl, P.K. (1999). *The scientist in the crib: Minds, brains, and how children learn.* New York: William Morrow. Brings together a wealth of research on infant and toddler learning to show how, and how much, very young children know and parents naturally teach them. Includes chapters on the young child's knowledge about people, objects, and language and scientists' discoveries about the young brain and mind.

Piper, T. (1998). *Language and learning: The home and school years* (2nd ed.). Upper Saddle River, NJ: Prentice-Hall. Discusses how children learn language, how it is taught, and how the two are sometimes at odds. Traces language acquisition from birth through the school years, using experiences of a number of different children to illustrate stages and sequences of development.

Tryphon, A., & Vonèche, J. (Eds.). (1996). *The social genesis of thought.* Washington DC: Psychology Press. Presents the outcome of a week-long meeting of child development experts held at the Jean Piaget Archives in Geneva to debate Piaget's and Vygotsky's theories.

PUZZLE 5.1 TERM REVIEW

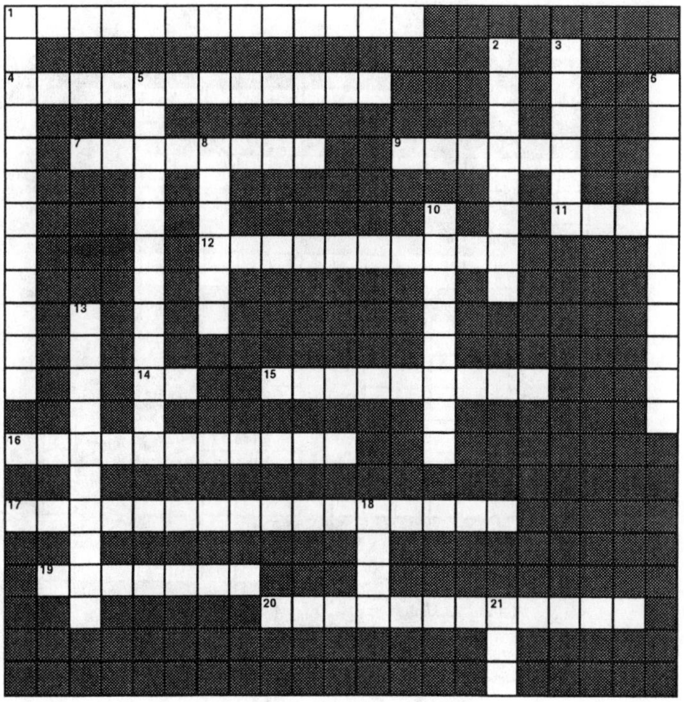

Across

1 Creating new schemes and adjusting old ones
4 Piaget's first stage
7 Sensory _____: where information is held briefly before decaying or moving on
9 _____ permanence: the understanding that objects continue to exist when out of sight
11 _____-term memory: permanent knowledge base; unlimited capacity
12 _____ style: vocabulary consisting of many pronouns and social formulas
14 A score permitting comparison of intelligence (abbr.)
15 Physical _____: the causal action one object exerts on another through contact
16 _____ behavior: deliberately combing schemes to solve a problem
17 _____ memory: recall of one-time events that are long lasting because they are imbued with personal memory
19 ____ memory: where a limited amount of information is worked on
20 Type of play involving pretend (2 words)

Down

1 Interpreting the external world in terms of current schemes
2 _____ imitation: copying the behavior of models not immediately present
3 _____ representation: an internal image of an absent object or past event
5 The internal rearranging and linking of schemes
6 Memory involving noticing whether a stimulus is similar to one previously experienced
8 Specific structure, or organized way of making sense of experience
10 A means of building schemes by repeating chance events, called the _____ reaction
13 Building schemes and adjusting old ones
18 Checklist for gathering information about the quality of children's home lives (abbr.)
21 An innate, biological system which allows children to speak in a rule-oriented fashion (abbr.)

PUZZLE 5.2 TERM REVIEW

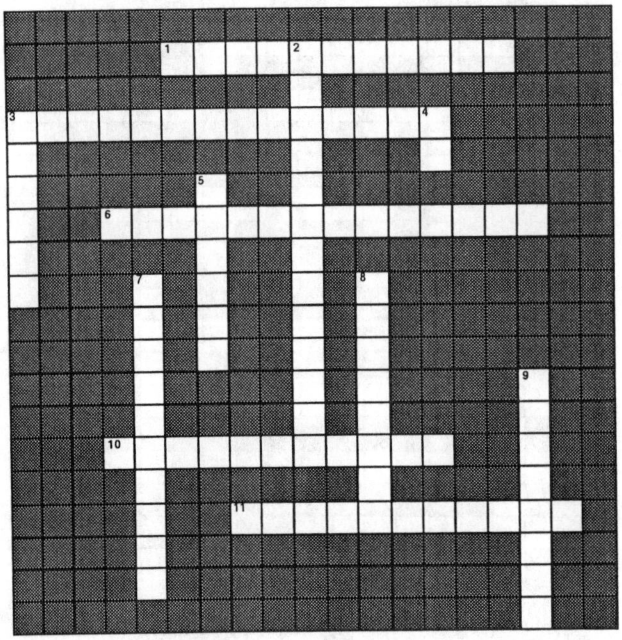

Across

1 Developmentally _____ practice: standards devised by the National Association for the Education of Young Children

3 A form of speech used to speak to infants and toddlers (2 words)

6 _____: applying a word too narrowly

10 Style of early language in which toddlers mainly label objects

11 _____ speech: two-word utterances that often leave out smaller and less important words

Down

2 _____: applying a word too broadly

3 Vowel-like noises produced by babies

4 Conservative label for infant test scores (abbr.)

5 Memory involving a stimulus not present

7 Mental ____ increases the efficiency of thinking.

8 Repetition of consonant-vowel combinations in a long string

9 Zone of _____ development: Vygotskian concept that refers to a range of tasks that the child cannot yet handle alone but can do with some adult help

SELF-TEST

1. According to Piaget, the first stage of cognitive development is the (148)
   a. preoperational stage.
   b. sensorimotor stage.
   c. concept stage.
   d. scheme development stage.

2. According to Piaget, specific structures which change with age and represent the child's organized way of making sense of experience are called (148)
   a. semi-structures.
   b. organismic values.
   c. schemes.
   d. sensori-structures.

3. Two-year-old Janet was fascinated by her older brother John's hamster. "See," John said, "hamster, hamster." Janet looked at him and replied, "Cat." Janet ________________________ the hamster into her scheme for cat. (148-149)
   a. assimilated
   b. accommodated
   c. organized
   d. operationalized

4. In Piaget's theory, a means of building schemes in which infants try to repeat a chance event caused by their own motor activity is called (149)
   a. accommodation.
   b. the circular reaction.
   c. the motoric reaction.
   d. a reflexive reaction.

5. Piaget described babies' behavior between birth and one month of age as (149)
   a. primary circular reactions.
   b. reflexive.
   c. scheme coordination.
   d. operantly conditioned.

6. During Substage 4, if an object is moved from one hiding place to a second hiding place, babies will look only in the first hiding place. This is called (151)
   a. an object permanence fault.
   b. goal-directed action.
   c. the AB search error.
   d. an obstacle course problem.

7. Children repeat behaviors with variation at about age (151)
   a. 3 to 5 months.
   b. 6 to 9 months.
   c. 10 to 12 months.
   d. 12 to 18 months.

8. The ability to remember and copy the behavior of models who are not immediately present is known as (152)
   a. deferred imitation.
   b. delayed imitation.
   c. functional imitation.
   d. make-believe imitation.

9. Recent research concerning object permanence, or the understanding that objects continue to exist when out of sight, suggests that it (152-153)
   a. develops around age 8-12 months, as Piaget theorized.
   b. exists as early as 3 1/2 months, but no earlier.
   c. may be prewired from the start, although not expressed in terms of search strategies until later.
   d. may be prewired from the start, because young infants demonstrate attempts to find hidden objects.

10. Information-processing and Piagetian approaches (156)
    a. both focus on the child as an active interpreter of information.
    b. are both explicit and precise.
    c. share the objective of describing in detail what children do, and in what order, when faced with a task.
    d. investigate the same aspects of thinking.

11. In information processing, procedures that operate on and transform information, increasing the efficiency of thinking and the chances that information will be retained, are called (156)
    a. mental strategies.
    b. sensory registers.
    c. flow charts.
    d. zones of proximal development.

12. The capacity of long-term memory is (157)
    a. about 7 to 9 pieces of information.
    b. limitless.
    c. several hours.
    d. two weeks.

13. What do information-processing theorists believe is largely responsible for changes in children's thinking? (157)
    a. the continued development of new brain structures
    b. an increase in capacity due to brain maturation
    c. improvements in strategies, such as attending and categorizing
    d. the development of conscious long-term memory

14. Habituation studies show that the infant's memory is (157)
    a. able to retain information for several hours by the end of the first year.
    b. equivalent to an adult's by three months of age.
    c. able to retain recognition of human faces for a few weeks by the end of the first year.
    d. virtually nonexistent until the end of the first year.

15. Recognition memory (157)
   a. is a form of recall memory.
   b. appears to be a conscious, deliberate activity.
   c. is the simplest form of retrieval.
   d. is more challenging than recall memory.

16. The greatest drawback of the information-processing perspective is that (160)
   a. its elements are not easily combined into a comprehensive theory of cognitive development.
   b. it explains cognitive development in terms of discrete stages.
   c. it relies heavily on interactive experiences.
   d. it overlooks the infant's ability to categorize stimuli.

17. Research which showed that children engaged in twice as much make-believe play when they were with their mothers than when they were alone suggests that (161)
   a. make-believe play is not discovered independently by toddlers.
   b. young children are too insecure to engage in make-believe play alone.
   c. mothers direct play too much, thereby inhibiting cognitive development.
   d. mothers are not appropriate playmates for their toddlers.

18. A child with an IQ of 100 would have a percentile rank of (162)
   a. 25.
   b. 50.
   c. 85.
   d. 98.

19. High-quality child care has been shown to (165-166)
   a. have a negative impact on social skills.
   b. reduce the negative impact of a stressed, poverty-stricken home life.
   c. have no impact on children's cognitive, emotional, and social competence.
   d. increase children's fear of separation.

20. Which of the following provides evidence that children are biologically primed to acquire language? (167)
   a. Children's first word combinations do not appear to follow grammatical rules.
   b. There is as yet no evidence of a single system of grammar.
   c. Children all over the world reach major milestones of language development in a similar sequence.
   d. Children's progress in mastering many sentence constructions is steady and gradual.

21. By the end of the first year, babies use ____________ to intentionally influence the behavior of others. (169)
   a. babbling sounds
   b. extensions
   c. preverbal gestures
   d. nativism

22. Children's knowledge of categorical relations would lead us to predict that they would be most likely to overextend the word *doggie* to (169-170)
    a. other animals.
    b. any living thing.
    c. things that make loud noises.
    d. nearly anything; their errors are random.

23. Toddlers use unique styles of early language learning. Those whose early vocabularies consist mainly of words referring to objects have a(n): (170)
    a. expressive style.
    b. referential style.
    c. concrete style.
    d. episodic style.

24. Child-directed speech works effectively to (171)
    a. discipline children.
    b. encourage active play.
    c. check and enhance comprehension.
    d. teach language formally.

25. Conversational give-and-take between parent and toddler is (172)
    a. one of the best predictors of early language development and academic competence.
    b. unnecessary; children develop language naturally without environmental support.
    c. rare, since most toddlers cannot yet converse.
    d. too difficult for young children.

CHAPTER 6

EMOTIONAL AND SOCIAL DEVELOPMENT IN INFANCY AND TODDLERHOOD

BRIEF CHAPTER SUMMARY

Erikson's psychoanalytic theories provide an overview of the emotional and social tasks of infancy and toddlerhood. According to Erikson, trust and autonomy grow out of warm, supportive parenting and reasonable expectations for impulse control during the second year.

Emotions play an important role in the organization of relationships with caregivers, exploration of the environment, and discovery of the self. Infants' ability to express basic emotions and respond to the emotions of others expands over the first year. Researchers agree that signs of emotions such as happiness, interest, surprise, fear, anger, and sadness are present early in infancy and, over time, become well-organized signals. As toddlers become more self-aware, self-conscious emotions such as shame, embarrassment, and pride begin to emerge. The ability to self-regulate emotions improves with brain maturation, gains in cognition and language, and sensitive child rearing.

Children's unique temperamental styles are apparent in early infancy. A growing body of research explores temperament, including stability, its biological roots, and its interaction with child-rearing practices.

Ethological theory is the most widely accepted view of the development of the infant–caregiver relationship. According to this perspective, attachment evolved over the history of our species to promote survival. Well-coordinated, positive emotional communication between caregiver and baby supports secure attachment, while insensitive caregiving is linked to attachment insecurity. Infants form attachment bonds with a variety of familiar people, including mothers, fathers, siblings, grandparents, and substitute caregivers.

During the first 2 years the knowledge of the self as a separate, permanent identity emerges. The first aspect to emerge is the I-self, followed in the second year by the me-self. Empathy, the ability to categorize the self (according to age, sex, and goodness and badness), compliance, and self-control are all by products of toddlers' emerging sense of self.

LEARNING OBJECTIVES

After reading this chapter, you should be able to:

6.1. Explain Erikson's stages of basic trust versus mistrust and autonomy versus shame and doubt, noting major personality changes. (178-179)

6.2. Describe the development of happiness, anger, and fear over the first year, noting the adaptive functions of each. (179-183)

6.3. Summarize changes in infants' ability to understand and respond to the emotions of others. (183)

6.4. Explain why self-conscious emotions emerge during the second year, and indicate their role in development. (183-184)

6.5. Trace the development of emotional self-regulation during the first 2 years. (184)

6.6. Describe the meaning of temperament, the ways in which it is measured, and major temperamental styles. (184-188)

6.7. Discuss the role of heredity and environment in the stability of temperament, including the goodness-of-fit model. (188-189)

6.8. Describe the unique features of ethological theory of attachment and the development of attachment during the first 2 years. (190-191)

6.9. Describe the Strange Situation, the four attachment patterns assessed by it, and cultural variations in infants' reactions to this procedure. (191-192)

6.10. Discuss factors that affect attachment security. (192-195)

6.11. Compare fathers' and mothers' attachment relationships with infants, and note factors that affect early sibling relationships. (195-197)

6.12. Describe and interpret the relationship between secure attachment in infancy and cognitive, emotional, and social competence in childhood. (198)

6.13. Trace the emergence of self-awareness, and explain its role in the development of empathy, categorizing the self, and self-control. (198-200)

STUDY QUESTIONS

Erikson's Theory of Infant and Toddler Personality

1. Erikson expanded and enriched Freud's view of development during the oral stage. According to Erikson, a healthy outcome during infancy does not depend on the ______________ of food or stimulation offered, but the ______________ of the caregiver's behavior. (178)

2. According to Erikson, the basic psychological conflict of the first year is ______________________________________. Under what conditions is this conflict resolved on the positive side? (178-179)

____

3. How did Erikson expand and enrich Freud's view of development during the anal stage? (179)

____

____

4. According to Erikson, the basic psychological conflict of the second year is ____. Under what conditions is this conflict resolved on the positive side? (179)

____

____

Emotional Development

1. Define and provide examples of *basic emotions.* (180)

____

____

2. True or False: Most researchers agree that almost all basic emotions are evident in early infancy. (180)

3. Children of depressed parents are _____ to _____ times more likely to develop behavior problems than children of nondepressed parents. (180)

4. Explain how the parent-child relationship is affected by maternal depression. (180-181)

____

____

5. Cite two ways in which infants' expressions of happiness contribute to development. (182)

A. ____

B. ____

6. List stimuli that trigger infants' smiles at the following ages: (182)

Newborn: ____

____

End of 1st month: ____

6 to 10 weeks: ____

7. Babies first laugh around __________ months of age. Laughter reflects ____________ ____ than does smiling. (182)

8. Expressions of anger gradually increase in frequency and intensity from _______ months into the second year. (182)

9. The baby's most frequent expression of fear is to _______________________, a response called ____________________________.

Cite two factors that influence this response. (182)

A. ________________________________ B. ________________________________

10. Explain the significance of the rise in anger and fear at the end of the first year. (182)

Anger: ____

____

Fear: ____

____

11. What cognitive abilities influence the development of infants' angry and fearful reactions? (182-183)

Anger: ____

Fear: ____

12. Explain how culture can influence the development of emotion in infants. (183)

____

____

13. Define *social referencing,* and explain its role in infant development. (183)
Definition:____
____
Functions: ____
____

14. Self-conscious emotions are so named because they involve injury to or enhancement of our ______________________. Provide some examples of such emotions. (183)
____

15. Besides ______________________, self-conscious emotions require ____________ __________________ in when to feel proud, ashamed, guilty, etc. (183)

16. Provide three examples of cultural differences in the self-conscious emotions children are encouraged to feel in response to various situations. (184)
A. ____
B. ____
C. ____

17. _________________________________ refers to the strategies we use to adjust our emotional state to a comfortable level of intensity so that we can accomplish our goals. List two examples from your own experience. (184)
____
____

18. True or False: Parents foster the development of emotional regulation in the early months by engaging in stimulating interaction while adjusting the pace of their own behavior so that the infant does not become distressed. (184)

19. As caregivers help infants regulate their emotions, they provide lessons in ______________________________ ways of expressing feelings. Briefly describe how American middle-class parents teach their infants about appropriate expressions of positive and negative emotions. (184)
____
____

20. By the end of the second year, ______________________________________ leads to more effective ways of regulating emotions. (184)

Temperament and Development

1. Define *temperament.* (185)
____
____

2. Match each of the following temperamental types identified in the New York Longitudinal Study with its appropriate description: (185)

____	Inactive, low-key reactions to stimuli, negative in mood, adjusts slowly to new experiences	1. Easy child
____	Regular in daily routines, cheerful, and adapts easily to new experiences	2. Difficult child
____	Irregular in daily routines, slow to accept new experiences, tends to react negatively and intensely	3. Slow-to-warm-up child

3. True or False: All children in the New York Longitudinal Study fit one of the above temperamental patterns. (185)

4. List three ways in which temperament is measured. Indicate which one is used most often. (187)
A. ____________________
B. ____________________
C. ____________________

5. Parental reports are useful because of their ____________________ and parents' ____________________. However, parental information is also ____________________ and is only modestly related to observational measures. (187)

6. According to Kagan, individual differences in arousal of the ____________________ and the pattern of EEG waves in the frontal region of the ____________________ contribute to differing temperamental styles. (187)

7. Compare the following physiological reactions of shy, inhibited babies with social babies to highly stimulating, unfamiliar experiences. (186)
Heart rate: ____________________
Cortisol: ____________________
Pupil dilation and blood pressure: ____________________

8. Temperamental stability from one age period to the next is generally ____________ to ____________. (188)

9. True or False: Biologically based temperamental traits cannot be modified by experience. (188)

10. Research shows that identical twins (are/are not) more similar than fraternal twins in temperament and personality. (188)

11. Describe some ethnic and sex differences in early temperament that appear to support a role for heredity. (188)
____
____
____

12. Some differences in early temperament are encouraged and maintained by ____________________. A similar process contributes to sex differences in temperament. (188)

13. In families with several children, parents often emphasize ______________ ______________. This effect (increases/decreases) temperamental differences among siblings. (188-189)

14. Explain the concept of *goodness-of-fit.* (189)

____

____

____

15. Briefly describe the pattern of parent-child interaction that often develops in the case of difficult children in Western society. (189)

____

____

____

16. Provide an example which illustrates the idea that goodness-of-fit depends in part on cultural values. (189)

____

____

____

Development of Attachment

1. ______________ is the strong affectional tie we feel for special people in our lives. (190)

2. True or False: Psychoanalytic theory and behaviorism both emphasize the importance of feeding in the development of attachment. (190)

3. Provide three examples of evidence that the development of attachment is not dependent on feeding. (190)

A. ______________

____

B. ______________

C. ______________

4. Ethological theory of attachment is based on the belief that the human infant is born with a set of innate behaviors which increase the chances that it will be ______________ ______________. (190)

5. Match the following stages of attachment from Bowlby's ethological theory with the appropriate descriptions. (190-191)

____ Cognitive improvements permit the prediction of parents' coming and going.
____ Birth to 6 weeks
____ Babies respond differently to a familiar caregiver than to a stranger, but do not protest when separated.
____ 6 weeks to 6-8 months
____ Babies are not yet attached, since they don't respond differently to unfamiliar adults.
____ 6-8 months to 18 months-2 years
____ Separation anxiety is displayed.
____ 18 months to 2 years and beyond
____ Built-in behaviors help bring newborns into close contact with other humans.
____ Familiar caregivers are used as a secure base from which to explore.
____ Children start to negotiate with caregivers using requests and persuasion.

1. Preattachment phase
2. "Attachment in the making" phase
3. Phase of "clearcut" attachment
4. Formation of a reciprocal relationship

6. According to Bowlby, children construct an *internal working model* out of experiences during the first 2 years. Define and explain this term. (191)

____
____
____

7. The reasoning behind the Strange Situation was that securely attached infants should use the parent as a ______________________ from which to explore, and when the parent leaves, the child should show ________________________ and find a strange adult (more/less) comforting. (191)

8. Match each of the following attachment classifications assessed by the Strange Situation with its appropriate description. (191-192)

____ Before separation, these infants seek closeness to the parent and fail to explore. When she returns, they display angry behavior and may continue to cry.
____ Before separation, these infants use the parent as a base from which to explore. They are upset by the parent's absence, and when she returns, they seek contact and are easily comforted.
____ Before separation, these infants seem unresponsive to the parent. When she leaves, they react to the stranger in much the same way as to the parent. Upon her return, they are slow to greet her.
____ When the parent returns, these infants show confused, contradictory behaviors, such as looking away while being held.

1. Secure
2. Avoidant
3. Resistant
4. Disorganized/disoriented

9. Due to cultural differences, _____________ babies show more avoidant attachment and _______________ infants show more resistant attachment than do American babies. Nevertheless, the ________________ pattern remains the most common pattern in all societies studied. (192)

10. According to Spitz, institutionalized babies had emotional difficulties because they were prevented from ______________________________________. Fully normal development may depend on forming close bonds with caregivers during the __________ __________ years of life. (193)

11. A special pattern of mother–infant communication called, _______________ ____________________, separates the experiences of securely attached from insecurely attached babies. In this type of interaction, the caregiver responds to infant signals in a __________________________ fashion, and both partners match positive ___________ ____________________ states. (193)

12. True or False: Interaction between securely attached babies and their mothers is perfectly harmonious the majority of the time. (193-194)

13. Compared to securely attached infants, avoidant infants tend to receive (overstimulating/unresponsive) care, while resistant infants tend to receive (overstimulating/unresponsive) care. (194)

14. Among maltreated infants and infants of depressed mothers, ___________________ _________________________________ attachment is especially high. (191)

15. When parents have the _____________________________ to care for a baby with special needs, and when the infant is _____________________, at-risk newborns fare well in the development of attachment. (194)

16. The influence of temperament and other infant characteristics on attachment security may depend on _____________________. (194)

17. True or False: When families experience major life changes, insecure attachment is especially high. (194)

18. Mothers who show ____________________________________ in discussing their childhoods tend to have securely attached infants. Mothers who dismiss the importance of _______________ relationships or describe them in _________________________ ways usually have insecurely attached infants. (195)

19. How does the rate of attachment insecurity among child-care infants compare to that of home-reared babies and children in industrialized countries around the world? (196-197)

Home-reared babies: ____

Children in industrialized nations: ____

20. List five factors that may help explain why some child-care infants may be more likely to display insecure attachment. (196-197)

A. ____

B. ____

C. ____

D. ____

E. ____

21. In many cultures, including that of the United States, mothers devote more time to _______________ and ____________________, while fathers devote more time to ____________________ interaction. (195)

22. True or False: Employed mothers tend to engage in more playful stimulation of their babies than do unemployed mothers. (195)

23. True or False: A cooperative and intimate relationship between husband and wife supports the involvement of both parents, but especially fathers, with their babies. (195)

24. When a new baby arrives, how is a preschool sibling likely to behave? Include both negative and positive reactions. (196-197)
Negative:____
____
Positive: ____
____

25. List two factors that affect the quality of early sibling relationships. (197)
A. ________________________________ B. ________________________________

26. What aspects of development are related to attachment security in infancy? (198)
____
____

27. True or False: Secure infants always show more favorable development than insecure infants. (198)

28. What determines whether attachment insecurity is linked to later problems? Explain. (198)
____
____
____

Self-Development During the First Two Years

1. _________ is the earliest aspect of the self to emerge. Describe how it emerges. (199)
____
____
____

2. During the second year, the __________ begins to emerge. Cite three behaviors indicating that by two years of age, recognition of the self is well established. (199)
A. ____
B. ____
C. ____

3. True or False: The development of the self is fostered by sensitive caregiving. (199)

4. Self-awareness leads to the first signs of ________________ as toddlers give to others what they themselves find most comforting. (199)

5. During toddlerhood, children categorize themselves and others on the basis of _______________, _______________, _______________, and even _______________. They use this knowledge to _______________ _______________. (200)

6. Define self-control, and list two milestones of development that support it. (200)
Definition: _______________
____

A. _______________
B. _______________

7. True or False: Among toddlers who experience warm, sensitive caregiving and reasonable expectations for mature behavior, compliance is more common than opposition. (200)

8. Around _____ months, the capacity for self-control appears and improves steadily into early childhood. (200)

9. Still, toddlers' control over their own actions is fragile, depending on constant _______________ and _______________ by parents. (200)

ASK YOURSELF . . .

Derek's mother fed him in a warm and loving manner during the first year, but when he became a toddler, she kept him in a playpen for many hours because he got into so much mischief while exploring freely. Use Erikson's theory to evaluate Derek's early experiences. (see text page 179)

____
____
____
____

Return to Chapter 1, page 18, and review Erikson's stages of psychosocial development. Select two stages beyond infancy and toddlerhood. Explain how trust and autonomy help the developing person resolve these later psychological conflicts positively. (see text pages 178-179)

____
____
____
____

Dana is planning to meet her 10-month-old niece Laureen for the first time. How should Dana expect Laureen to react? How would you advise Dana to go about establishing a positive relationship with Laureen? (see text page 182)

____
____
____
____

One of Byron's favorite games was dancing with his mother while she sang "Old MacDonald," clapping his hands and stepping from side to side. At 14 months, Byron danced joyfully as Beth and Felicia watched. At 20 months, he began to show signs of embarrassment—smiling, looking away, and covering his eyes with his hands. What explains this change in Byron's emotional reaction? (see text pages 183-184)

Compare findings on the development of emotional self-regulation with the consequences of maternal depression for children's development (see pages 180 and 181). Do children of depressed mothers have lasting difficulties with emotional self-regulation? Explain. (see text page 184)

Rachel, like many other Asian infants, is calm and easily soothed when upset. What factors contribute to her temperamental style? (see text pages 188-189)

At 18 months, highly active Byron climbed out of his high chair long before his meal was finished. Exasperated with Byron's behavior, his father made him sit at the table until he had eaten all his food. Soon Byron's behavior escalated into a full-blown tantrum. Using the concept of goodness-of-fit, suggest another way of handling Byron. (see text page 189)

Citing evidence on inhibited, or shy, children and uninhibited, or sociable, children, show how biology, parenting practices, and culture jointly influence the development of temperament. (see text pages 186-187)

Recall from Chapter 5 that Lisa tended to overwhelm Byron with questions and instructions. How would you expect Byron to respond in the Strange Situation? Explain your answer. (see text page 191-192)

As a child, Suzanne had a conflict-ridden relationship with her parents and felt angry and displaced when her sister was born. Will Suzanne's negative childhood experiences necessarily undermine her capacity to build a secure attachment with her infant daughter? Explain. (see text pages 194-195)

Maggy works full time and leaves her 14-month-old son Vincent at a child-care center. When she arrives to pick him up at the end of the day, Vincent keeps on playing and hardly notices her. Maggy wonders whether Vincent is securely attached. Is Maggy's concern warranted? (see text page 196)

Nine-month-old Harry turned his cup upside down and spilled juice all over the tray of his high chair. His mother said sharply, "Harry, put your cup back the right way!" Can Harry comply with his mother's request? Why or why not? (see text page 200)

Why is the development of compliance and self-control crucial for later cognitive and social development? (see text page 200)

SUGGESTED READINGS

Bretherton, J. (1994). The origins of attachment theory: John Bowlby and Mary Ainsworth. In R. D. Park, P. A. Ornstein, J. R. Rieser, & C. Zahn-Waxler (Eds.), *A century of developmental psychology* (pp. 431–472). Washington, DC: American Psychological Association. Summarizes John Bowlby's and Mary Ainsworth's ethological approach to attachment.

Campbell, A., & Muncer, S. (Eds.) (1997). *The social child.* Washington, DC: Psychology Press. Edited book that provides survey of most pressing issues in field of social development. Chapters cover information on cross-cultural findings, behavioral genetics, social cognition, family influences, and effects of the media on children's behavior.

Kagan, J. (1994). *Galen's prophecy: Temperament in human nature.* New York: Harper Collins. One of the foremost researchers on infant temperament provides a readable conceptualization of how it may affect individuals throughout the lifespan. The findings are based on 15 years of research with young children and suggest that temperamental factors may be influential in various aspects of development.

Meins, E. (1997). *Security of attachment and the social development of cognition.* Washington, DC: Psychology Press. Investigates how children's attachment styles are associated with various aspects of cognitive development throughout the preschool years. Accounts for individual differences using principles from both Bowlby's and Vygotsky's theories.

Taylor, R. D., & Wang, M. C. (Eds.) (1997). *Social and emotional adjustment and family relations in ethnic minority families.* Mahwah, NJ: Erlbaum. Collection of essays addressing issues related to the intersection of family relationships and several contexts for social and emotional development of ethnic minority adolescents.

PUZZLE 6.1 TERM REVIEW

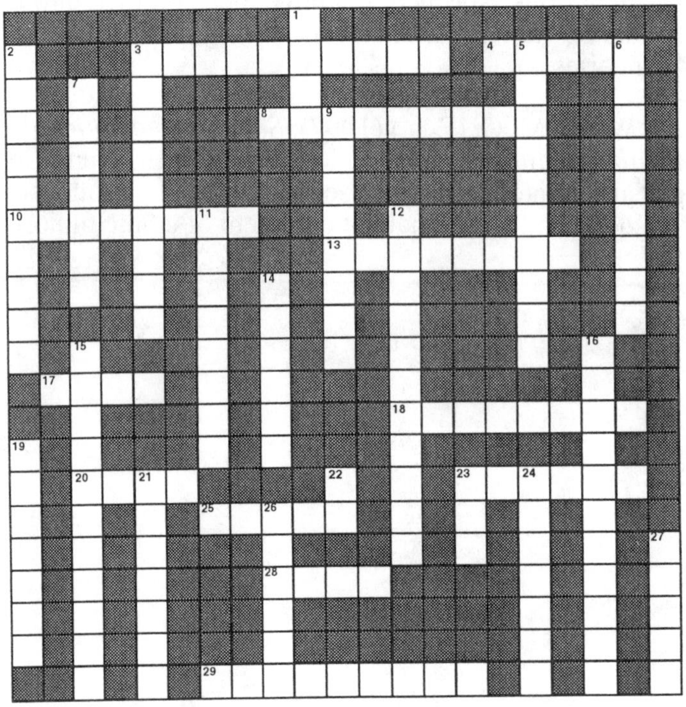

Across

3 _____ anxiety: an infant becomes upset when the adult whom she come to rely on leaves
4 _____ emotions can be directly inferred from facial expressions
8 _____ attachment: remains close before separation; is angry upon return
10 ____ working model: serves as a guide for future close relationships
13 Positive outcome of Erikson's psychological conflict of toddlerhood
17 Emotional self-_____: strategies for adjusting emotional state to a comfortable level
18 _____-of-fit: effective match between child rearing and temperament
20 _____ temperament: regular routines, cheerfulness, and adaptability
23 _____ attachment: distressed by separation; easily comforted upon return
25 Social _____: appears around 6-10 weeks
28 _____-to-warm-up temperament: inactive, shows mild, low-key reactions to new experiences
29 _____ temperament: irregular routines and negative, intense reactions

Down

1 Infants who explore the environment then return to their mothers for emotional support are using their mother as a secure ______
2 Voluntary obedience to adult requests
3 Interactional _____: sensitively tuned "emotional dance"
5 Infants' strong affectional tie to familiar caregivers
6 Self-_____ emotions: second, higher-order set of feelings, including shame, guilt, envy
7 Ability to understand and respond sympathetically to others' feelings
9 Uninhibited, or _____, children display positive emotion to and approach novel objects and people
11 _____ attachment: not distressed by separation; avoids parent upon return
12 _____ theory: most widely accepted perspective on attachment
14 _____ referencing: relying on a trusted person's emotional reaction to decide how to respond
15 Stable individual differences in quality and intensity of emotional reaction
16 Disorganized-_____ attachment: at reunion infants show a variety of confused behaviors
19 Stranger _____: fear of unfamiliar adults
21 _____ Situation: assesses quality of the attachment bond
22 _____-self: second aspect of the self to emerge
23 Inhibited, or _____, children react negatively to and withdraw from novel stimuli
24 Self-_____: the capacity to resist an impulse to engage in socially disapproved behavior
26 ___ __: earliest aspect of the self to emerge
27 Positive outcome of Erikson's psychological conflict of infancy

SELF-TEST

1. Erikson's stage of basic trust versus mistrust builds on Freud's _________ stage. (178)
   a. ego
   b. symbiotic
   c. anal
   d. oral

2. According to Erikson, successful resolution of the conflict of autonomy versus shame and doubt requires (179)
   a. parents who are very controlling and able to set rigid limits.
   b. parents who exercise very little control, leaving their children free to explore.
   c. parents who provide suitable guidance while allowing their children to make reasonable choices.
   d. that the child be successfully toilet trained by the age of two.

3. Emotions that can be directly inferred from facial expressions and that are present during the early weeks of life are called (180)
   a. basic emotions.
   b. learned responses.
   c. temperament.
   d. phobias.

4. Depressed parents (180-181)
   a. are more likely to have happy marriages.
   b. have children with a more positive, pleasant mood.
   c. use consistent discipline.
   d. use inconsistent discipline.

5. The social smile typically appears (182)
   a. at birth.
   b. between 6 and 10 weeks.
   c. at 3 to 4 months.
   d. between twelve to 18 months.

6. Infants' expression of fear toward unfamiliar adults during the second half of the first year is called (182)
   a. separation anxiety.
   b. the strange situation.
   c. stranger anxiety.
   d. stranger fear.

7. Babies who actively seek information about a trusted person's feelings are engaging in (183)
   a. self-conscious emotions.
   b. social referencing.
   c. basic emotions.
   d. social dynamics.

8. Self-conscious emotions appear in the _____________, as the sense of self emerges and toddlers can combine separate emotions. (183)
   a. first year
   b. second year
   c. third year
   d. fourth year

9. The capacity to adjust one's emotional state to a comfortable level of intensity is known as (184)
   a. emotional self-regulation.
   b. social referencing.
   c. self-referencing.
   d. self-conscious emotional regulation.

10. Which of the following is true regarding temperament? (185-186)
    a. Shy, inhibited children are more likely to have dark hair and brown eyes.
    b. Genes make a modest contribution to shyness and sociability.
    c. Child-rearing practices do not influence temperament.
    d. Shy, inhibited children often have cognitive difficulties.

11. Research indicates that ___________________ in early temperament. (188)
    a. ethnic and sex differences exist
    b. shyness is not evident
    c. there are no cultural differences
    d. heredity plays no role

12. An effective match between child-rearing practices and the child's temperament that leads to favorable adjustment is known as (189)
    a. the caregiving continuum.
    b. attachment.
    c. goodness-of-fit.
    d. parenting temperament.

13. According to the _____________ theory of attachment, the baby desires closeness to the mother because her presence is paired with relief of hunger. (190)
    a. ethological
    b. psychoanalytic
    c. behaviorist
    d. ecological

14. Ethological theorists view attachment as resulting from (190)
    a. a secondary drive for affection.
    b. innate behaviors that elicit parental care.
    c. reinforced behavior.
    d. reduction of anxiety which leads to trust.

15. Which of the following lists the stages of ethological theory of attachment in the correct order? (190-191)
    a. attachment-in-the-making, reciprocal relationship, clearcut attachment
    b. clearcut attachment, attachment-in-the-making, reciprocal relationship
    c. preattachment, attachment-in-the-making, clearcut attachment
    d. preattachment, reciprocal relationship, clear attachment

16. A baby who, after being reunited with the mother, hits and pushes in anger is displaying which form of attachment? (191-192)
    a. avoidant
    b. resistant
    c. secure
    d. disorganized/disoriented

17. A sensitively tuned interaction in which the mother responds to infant signals in a well-timed, appropriate fashion and both partners match emotional states is called (193)
    a. interactional synchrony.
    b. maternal sensitivity.
    c. secure synchrony.
    d. reciprocity.

18. Which attachment pattern is especially high among maltreated infants? (194)
    a. avoidant
    b. resistant
    c. secure
    d. disorganized/disoriented

19. Which of the following qualities of child care does NOT foster attachment security? (196-197)
    a. small group sizes
    b. educated caregivers
    c. low caregiver-child ratios
    d. positive child-caregiver interactions

20. Fathers who are highly involved with their children tend to be (195)
    a. less assertive.
    b. less masculine.
    c. less confident in their identity.
    d. less gender-stereotyped in their beliefs.

21. When mothers and fathers play with infants, (195)
    a. they do so in a similar manner.
    b. fathers are more gentle because they fear hurting the baby.
    c. mothers engage in more exciting, physical play.
    d. fathers engage in more exciting, physical play.

22. According to Lamb, an infant who is insecurely attached in infancy may fare well later if (198)
    a. there are compensating affectional ties outside of the family.
    b. the infant is neglected rather than maltreated, and is not rejected in school.
    c. the parents are of high socioeconomic status.
    d. the infant is an only child.

23. A sense of self as a subject who is separate from but attends to and acts on objects and other people is called the (199)
    a. me-self
    b. empathic-self
    c. I-self
    d. separate-self

24. Compliance with adult requests and commands is (198)
    a. rare before 18 months.
    b. one of the first forms of self-control to emerge.
    c. detrimental to the self-concept.
    d. does not occur in the toddler years.

25. Self-awareness provides for all of the following EXCEPT (199-200)
    a. categorizing the self.
    b. self-control.
    c. compliance.
    d. secure attachment.

CHAPTER 7

PHYSICAL AND COGNITIVE DEVELOPMENT IN EARLY CHILDHOOD

BRIEF CHAPTER SUMMARY

While body growth slows during early childhood, the brain continues to grow faster than other parts of the body. Lateralization increases, and language development is supported by the more rapid growth of the left hemisphere. Myelinization continues, and connections between parts of the brain increase, supporting improvements in physical and cognitive skills. Physical growth is affected by heredity through the regulation of hormone production, but as in earlier periods, environment contributes considerably. Emotional deprivation and malnutrition can interfere with physical development, and illness can interact with malnutrition to undermine children's growth. However, the leading cause of childhood death is unintentional injury. Physical development during early childhood is exemplified by an explosion of gross and fine motor skills, influenced by a combination of hereditary and environmental factors.

Dramatic advances in mental representation occur during early childhood. Although Piaget's preoperational stage emphasizes cognitive limitations, research on the young child's theory of mind reveals that preschoolers use logical, reflective thought on familiar tasks. Still, Piaget's theory has had a powerful influence on education, promoting child-oriented teaching approaches. Vygotsky regarded language as the foundation for all higher cognitive processes. As adults and skilled peers provide children with verbal guidance on challenging tasks, children incorporate these dialogues into their own self-directed speech. A Vygotskian approach emphasizes assisted discovery, including verbal support and peer collaboration. Some of Vygotsky's ideas, like Piaget's, have been challenged, especially regarding the role verbal communication plays in developing children's thinking in different cultures.

A variety of information-processing skills improves during early childhood. Like adults, young children remember everyday events in a logical, well-organized fashion. Preschoolers develop a basic understanding of written symbols and arithmetic concepts through informal experiences. A stimulating environment, warm parenting, and reasonable demands for maturity predict mental development in early childhood. While at-risk children show long-term benefits from early intervention and high-quality child care, poor child care undermines the development of all children. Language development proceeds at a rapid pace in early childhood, supported by conversational give-and-take. By the end of the preschool years, children have an extensive vocabulary, use most grammatical constructions competently, and are effective conversationalists.

LEARNING OBJECTIVES

After reading this chapter, you should be able to:

7.1. Describe changes in body size, proportions, and skeletal maturity during early childhood, and discuss asynchronies in physical growth. (207-210)

7.2. Discuss brain development in early childhood, including lateralization and handedness and myelinization of the cerebellum, reticular formation, and corpus callosum. (210-212)

7.3. Describe the impact of heredity, emotional well-being, nutrition, and infectious disease on early childhood growth, and compare child health services in the United States with those of other Western nations. (212-216)

7.4. Summarize factors related to childhood injuries, and cite preventive measures. (216-217)

7.5. Cite major milestones along with individual and sex differences in gross and fine motor development in early childhood. (218-221)

7.6. Describe advances in mental representation during the preschool years, including changes in make-believe play. (221-223)

7.7. Describe the limitations of preoperational thought from Piaget's point of view. (223-224)

7.8. Discuss recent research on preoperational thought and its implications for the accuracy of Piaget's preoperational stage. (225-227)

7.9. Describe three educational principles derived from Piaget's theory. (227)

7.10. Contrast Piaget's and Vygotsky's views on the development and significance of children's private speech. (228)

7.11. Discuss applications of Vygotsky's theory to education, and summarize current challenges to his ideas. (229-230)

7.12. Describe the development of attention and memory during early childhood. (230-231)

7.13. Discuss preschoolers' awareness of an inner mental life, factors that support their early understanding, and limitations of their theory of mind. (231-232)

7.14. Trace the development of preschoolers' literacy and mathematical knowledge, and discuss ways to enhance their academic development. (232-234)

7.15. Describe the impact of home, preschool and child care, and educational television on mental development in early childhood. (235-239)

7.16. Trace the development of vocabulary, grammar, and conversational skills, and cite factors that support language learning in early childhood. (239-241)

STUDY QUESTIONS

Physical Development

Body Growth

1. On the average, children add ________ inches in height and about _____ pounds in weight each year. (208)

2. Between ages 2 and 6, approximately ____ new epiphyses, or ______________ ______________, are added to the skeleton. (208)

3. By the end of the preschool years, children start to lose their ___________________. The timing of this process is affected by _____________________ and ___________________ influences. (209)

4. Although growth is asynchronous, the term __________________________________ refers to changes in overall body size involving ___________ growth during infancy, _____________ gains in early and middle childhood, and ____________ growth again during adolescence. (209)

Brain Development

1. Preschoolers improve in a wide variety of skills as _______________________ and _____________________ of neural fibers continue. (210)

2. For most children, the left hemisphere shows a dramatic growth spurt between __________ years and then levels off. The right hemisphere develops (slowly/rapidly) throughout early and middle childhood. This helps explain the pattern of development of what two skills? (210)
A. ________________________________ B. ________________________________

3. Hand preference is stable by age ______. (210)

4. True or False: The ambidextrous abilities displayed by many left-handers suggest that their brains tend to be more strongly lateralized than those of right-handers. (211)

5. Research indicates that twins are more likely to (be the same/be different) in handedness, and that hand preference may be related to _________________ during the prenatal period. (211)

6. What probably explains the finding that left-handedness is more frequent among severely retarded and mentally ill people than it is in the general population? (211)
____
____

7. For each of the following brain structures, describe developmental changes in early childhood, and indicate their impact on children's physical and cognitive skills: (211-212)

Cerebellum

Changes: ____
____
Impact: ____

Reticular Formation

Changes: ____
____
Impact: ____

Corpus Callosum

Changes: ____
____
Impact: ____

Influences on Physical Growth and Health

1. Without growth hormone (GH), children reach an average mature height of only about ________________. (212)

2. Explain why caution should be exercised in prescribing GH for short, GH-normal children. (213)
____
____
____

3. Thyroid-stimulating hormone (TSH) stimulates the release of thyroxin, necessary for normal development of the ____________________________ and for GH to have its full impact. (212)

4. True or False: Unlike other age periods, emotional stress is not associated with illness and unintentional injuries in early childhood. (214)

5. List the cause and three characteristics of deprivation dwarfism. (214)
Cause: ____
Characteristics: ____
____

6. Why does appetite decline in early childhood? What is the possible adaptive value of preschoolers' wariness of new foods? (214)
A. ____
B. ____

7. True or False: Preschoolers require the same balance of foods that make up a healthy adult diet. (214)

8. Cite two factors that influence young children's food preferences. (214)
A. ______________________________
B. ______________________________

9. List the five most common dietary deficiencies during the preschool years. (214-215)
A. ______________ B. ______________
C. ______________ D. ______________
E. ______________

10. In well-nourished children, ordinary childhood illnesses have ______________ impact on physical growth. In undernourished children, when disease interacts with malnutrition, the consequences of physical growth can be ______________. (215)

11. Nearly 1 million children are saved annually as a result of ______________, that provides them with a glucose, salt, and water solution that replaces lost fluids. (215)

12. How do immigration rates in the United States compare to those of other industrialized nations? (215-216)

____
____
____
____

13. What is the leading cause of childhood mortality? ______________
List the three most common types during the early childhood years. (216)
A. ______________ B. ______________
C. ______________

14. Cite two child characteristics that increase the risk of injury. (216)
A. ______________ B. ______________

15. Briefly describe family characteristics related to injury. (216-217)

____
____
____

16. List three factors that probably contribute to the high childhood injury rate in the United States. (217)
A. ______________ B. ______________
C. ______________

17. A variety of programs based on ______________ have been effective in improving the safety practices of adults and children. (217)

Motor Development

1. Which principle that governed motor development in the first 2 years continues to operate in early childhood? (218)

____

2. As children's bodies become more streamlined in early childhood, their ______________________________ shifts, and ______________ improves. (218)

3. Match the following sets of gross motor developments with the ages at which they are typically acquired. (218)

____	Walks up stairs with alternating feet; Flexes upper body when jumping and hopping; Throws with slight involvement of upper body, still catches against chest; Pedals and steers tricycle	1. 2 to 3 years
____	Walks down stairs with alternating feet, gallops; Throws ball with transfer of weight on feet, catches with hands; Rides tricycle rapidly, steers smoothly	2. 3 to 4 years
____	Hurried walk changes to true run; Jumps, hops, throws, and catches with rigid upper body; Little steering	3. 4 to 5 years
____	Engages in true skipping; Displays mature throwing and catching style; Rides bicycle with training wheels	4. 5 to 6 years

4. To parents, fine motor development is most apparent in what two areas? (218)
A. ________________________________ B. ______________________________

5. Match the following sets of fine motor developments with the ages at which they are typically acquired. (2118-220)

____	Draws first tadpole image of a person; Copies vertical line and circle; Uses scissors; Fastens and unfastens large buttons	1. 2 to 3 years
____	Draws a person with six parts; Copies some numbers and words; Ties shoes; Uses knife	2. 3 to 4 years
____	Copies triangle, cross, and some letters; Cuts along line with scissors; Uses fork effectively	3. 4 to 5 years
____	Scribbles gradually become pictures; Puts on and removes simple items of clothing; Zips large zippers; Uses spoon effectively	4. 5 to 6 years

6. Children's drawings reflect the conventions of their __________________. An emphasis on artistic expression produces ________________________ drawings. (220)

7. Letter reversals are common until ___________________________. (220)

8. ____________________ is thought to contribute to the superior performance of ______________________________ children over Caucasian children in running and jumping. (220)

9. During early childhood, boys are slightly ahead of girls in skills that emphasize __________________________ and ____________________________, while girls are ahead in ______________________ skills and gross motor skills that require good ____________________________ combined with foot movement. (220)

10. ____________________ for boys to be skilled at gross motor activities and for girls to play quietly at fine motor activities appears to exaggerate (small/large) genetically based sex differences. (220-221)

11. True or False: Preschoolers exposed to formal lessons in motor skills are generally ahead in motor development. (220)

12. Explain how adults can influence preschoolers' motor progress. (220)

____

____

Cognitive Development

Piaget's Theory: The Preoperational Stage

1. As children move from the sensorimotor to the preoperational stage, the most obvious change is an extraordinary increase in ______________________________. (221)

2. According to Piaget, (language/sensorimotor activity) gives rise to representational thought. (221)

3. Over time, make-believe play becomes increasingly (detached from/related to) real-life conditions because children become better at imagining objects and events (with/without) support from the real world. (221-222)

4. Make-believe play gradually becomes less _____-centered. (222)

5. Finally, make-believe play includes increasingly complex ____________________ combinations. (222)

6. Make-believe play both reflects and contributes to children's ____________ and ____________ skills. Give one example of each skill. (222-223)
A.______________________________ B.______________________________

7. In the preoperational stage, young children are not capable of *operations.* Define this term. (223)

____

8. The most serious deficiency of preoperational thinking is ______________, or the lack of awareness of perspectives other than one's own. (223)

9. Describe the three-mountains task for assessing egocentrism. (223)

____

____

____

10. Explain the meaning of *conservation.* (223)

____

____

11. Match each of the following features of preoperational thought highlighted by preschoolers' inability to conserve with the appropriate description. (223-224)

____	Focus on one aspect of a situation to the neglect of other features	1. Perception-based
____	Cannot mentally go through a series of steps and then trace steps back to the starting point	2. Centration
____	Focus on momentary conditions rather than dynamic changes between them	3. States vs. transformations
____	Easily distracted by the appearance of objects	4. Irreversibility

12. Preschoolers' performance on Piaget's class inclusion problem illustrates their difficulty with ________________________________. (224)

13. True or False: When researchers provide children with a display containing familiar materials instead of Piaget's three-mountains task, they no longer respond in an egocentric fashion. (225)

14. Cite two examples of nonegocentric responses in young children's everyday interactions. (225)

A. ____

B. ____

15. True or False: Preschoolers' animistic responses result from inadequate knowledge about some objects, not from a rigid belief that inanimate objects are alive. (225)

16. Between 4 and 8 years of age, as familiarity with physical events and principles ____________________, children's magical beliefs ____________________. (226)

17. Contrary to Piaget's theory, illogical thought seems to only occur when preschoolers are presented with ______________________________________ or are given ______________________________________. (226)

18. True or False: An examination of preschoolers' everyday knowledge shows that the capacity to classify hierarchically is present in early childhood. (226)

19. Engaging in ____________________________ may help children refine their understanding of what is real and what is not real. (226-227)

20. Research findings on preoperational thought indicate that the attainment of logical operations is a(n) (gradual/abrupt) development. (227)

21. List and briefly describe three educational principles derived from Piaget's theory. (227)

A. ____

____

B. ____

____

C. ____

____

Vygotsky's Sociocultural Theory

1. Reflecting his belief that young children are unable to take the perspective of others, Piaget called children's self-directed speech ________________ speech. (228)

2. Vygotsky reasoned that children speak to themselves for ____________________ and ___________________________. He viewed such speech as a foundation for all ____. According to Vygotsky, with age self-directed speech becomes ___________________________. (228)

3. Because most research supports Vygotsky's perspective, children's speech-to-self is now referred to as __________________________. (228)

4. Describe the techniques of *assisted discovery* and *cooperative learning* in the context of a Vygotskian classroom. (229)
Assisted discovery:____
____
Cooperative learning: ____
____

5. Vygotsky viewed _________________________ as the ideal social context for fostering cognitive development in early childhood. (229)

6. True or False: Verbal communication may not be the only means, or even the most important one, through which children's thinking develops. (229)

Information Processing

1. The capacity to sustain attention (remains stable/improves) during early childhood. (230)

2. Preschoolers' attention becomes more _____________ with age, but when given detailed pictures, they fail to search thoroughly. (230)

3. Even in comparison to adults, preschoolers' recall memory is much (better/poorer) than their recognition memory. (231)

4. Explain why preschoolers are not very strategic memorizers. (231)
____

5. Like adults, preschoolers remember familiar experiences in terms of ______________________. These become more __________________ with age. (231)

6. Between ages ____ and ____, children figure out that ___________________ and __________________ determine behavior. By age ____, they understand that people can hold false beliefs. (231)

7. List some of the characteristics of *infantile autism.* (233)
____
____

8. Young children view the mind as a ________________ container for information, while older children view it as a(n) _______________ agent that selects and transforms information and affects how the world is perceived. (232)

9. True or False: Young preschoolers understand how written language is related to meaning in an adultlike fashion. (232)

10. List ways in which adults can help prepare children for the tasks involved in reading and writing. (232-234)

____

____

11. List five steps in the development of preschoolers' mathematical reasoning that correspond with the following general age ranges: (234)

Early preschool: ____

Between 2 and 3: ____

Between 3 and 4: ____

Between 4 and 5: ____

Late preschool: ____

____

12. True or False: Basic mathematical knowledge emerges universally around the world, but at different rates. (234)

Individual Differences in Mental Development

1. Intelligence tests (do/do not) sample all human abilities and performance (is/is not) affected by cultural and situational factors. (235)

2. Intelligence test scores are important because they _____________________ which is related to ____________________________ in industrialized societies. (235)

3. Summarize factors that the early childhood version of HOME has found to be related to preschoolers' intellectual development. (235-236)

____

____

____

____

4. In child-centered preschools, teachers provide a wide variety of activities, and most of the day is devoted to ___________________________. In academic preschools, the program is (more/less) structured and teaches early academics through _________________________________. (236)

5. Describe the differences, revealed by two recent studies, in children who attended child-centered preschools compared to those who attended teacher-directed. (236)

____

____

6. True or False: High quality preschool experiences have a larger impact on low-SES than middle-SES children. (236)

7. A typical Head Start program provides preschool as well as _____________ _____________ services. _______________ involvement is a central component. (236)

8. Describe the long-term impact of preschool intervention on low-SES children's development. (236-239)

____
____
____
____
____

9. In addition to teaching ____________________ skills and providing other supports to parents, researchers advocate expanding early intervention to include _______________________________________ goals for both parents and children and helping parents move out of poverty with _______________________, ___________________, and other social services. (238)

10. True or False: Despite the developmental benefits of Head Start, the program does not appear to result in substantial financial savings to society. (238)

11. Preschoolers exposed to poor-quality child care score lower on measures of _________________ and ___________________ skills. (237)

12. List four key ingredients of high-quality child care in early childhood, and describe their effects on caregiver behavior and children's development. (237)

A. _______________________________ B. _______________________________
C. _______________________________ D. _______________________________
Effects: ____

13. The average 2- to 6-year-old watches TV for __________ hours a day. (239)

14. Describe the strengths and limitations of the rapid-paced format of "Sesame Street" for teaching preschoolers basic academic knowledge. (239)

Strengths: ____
____
Limitations: ____
____

15. In heavy doses, television viewing may encourage _____________ thinking. (239)

Language Development

1. A process called ___________________ helps explain how preschoolers build up their vocabularies quickly, from an average of _______ words at age 2 to _______ at age 6. (239)

2. True or False: Modifiers that are related to one another in meaning are the easiest kind for children to learn. (239)

3. When children assume that words refer to entirely separate categories, they are applying the principle of ___________________________________. (239)

4. Children use clues in the structure of _______________ and also rely on ___________ clues to learn the meaning of new words. (240)

5. As early as age 2, children __________________________ in systematic ways to fill in for ones they have not learned and use __________________ to extend language meanings and communicate in vivid and memorable ways. (240)

6. True or False: All English-speaking children master grammatical markers in a regular sequence. (240)

7. By age _____, children have acquired many grammatical rules , and they apply them so consistently that they occasionally ____________________________ the rules to words that are exceptions. (240)

8. Briefly describe the predictable errors preschoolers make in forming questions and demonstrating their understanding of the passive voice. (240)
Questions:____
____
Passive voice: ____
____

9. True or False: By the end of the preschool years, children have mastered most of the grammatical constructions of their language. (240)

10. True or False: Preschoolers are not yet capable of interacting competently in face-to-face conversation. (240)

11. By age ____, children know a great deal about culturally accepted ways of adjusting their speech to fit the _________________ , __________________ , and _________________ status of their listener as well as how well they know them. (240)

12. Opportunities for conversational ______________________ with adults are consistently related to general measures of language progress. (241)

13. Describe two techniques adults use to promote language skills when talking to preschoolers. (241)
A. ____
B. ____

ASK YOURSELF . . .

After graduating from dental school, Norm entered the Peace Corps and was assigned to rural India. He found that many Indian children had extensive tooth decay. In contrast, the young patients of his dental school friends in the United States had much less. What factors probably account for this difference? (see text page 209)

____

Crystal has a left-handed cousin who is mentally retarded. Recently she noticed that her 2-year-old daughter Shana is also left-handed. Crystal has heard that left-handedness is a sign of developmental problems, so she is worried about Shana. How would you respond to Crystal's concern? (see text page 211)

Explain why brain growth in early childhood involves not only an increase in neural connections but also a loss of synapses and cell death? Which assumption of the lifespan perspective do these processes illustrate? (If you need to review, see Chapter 1, page 9, and Chapter 4, page 120) (210-212)

One day Leslie prepared a new snack to serve at preschool: celery stuffed with ricotta cheese and pineapple. The first time she served it, few of the children touched it. How can Leslie encourage her pupils to accept the snack? What should she avoid doing? (see text page 214-215)

Using ecological systems theory (see Chapter 1), suggest ways to reduce childhood injuries by intervening in the microsystem, mesosystem, and macrosystem. (see text pages 25-27, 216-217)

Microsystem:

Mesosystem:

Macrosystem:

Using research on malnutrition or unintentional injuries, show how physical growth and health in early childhood result from continuous and complex interplay between heredity and environment. (214-217)

Mabel and Chad want to do everything they can do to support their 3-year-old daughter's motor progress and to foster an early, positive attitude toward athletics that will persist into the middle childhood and adolescence. What advice would you give them? (see text page 221)

Five-year-old Heather draws elaborate pictures at preschool, whereas Michael's drawings are limited to tadpole shapes with little detail. What factors might contribute to the gap in drawing competence between Heather and Michael? (see text pages 220-221)

Recently, 2-year-old Brooke's father shaved off his thick beard and mustache. When Brooke saw him, she was very upset. Using Piaget's theory, explain why Brooke was distressed by her father's new appearance. (see text page 223)

One weekend, 4-year-old Will went fishing with his family. When his father asked, "Why do you think the river is flowing along?" Will responded, "Because it's alive and wants to." Yet at home, Will understands very well that his tricycle isn't alive and can't move by itself. What explains this contradiction in Will's reasoning? (see text page 226)

Tanisha sees her 5-year-old son Toby talking out loud to himself while he plays. She wonders whether she should discourage this behavior. Use Vygotsky's theory to explain why Toby talks to himself. How would you advise Tanisha? (see text page 228)

According to Vygotsky, what ingredients are necessary at any age to promote learning and development? (see text page 229)

Piaget believed that preschoolers' egocentrism prevents them from reflecting on their own mental activities. What evidence in the preceding sections contradicts this assumption? (see text pages 231-232)

Lena wonders why Gregor's preschool teacher permits him to spend so much time playing instead of teaching him academic skills. Gregor's teacher responds, "I *am* teaching him academics—through play. It's the best way to ensure that he will become a competent, self-confident reader and writer." Explain what she means. (see text pages 232-234)

Senator Smith heard that IQ gains resulting from Head Start do not last, so he plans to vote against funding for the program. Write a letter to Senator Smith explaining why he should support Head Start and explaining what aspects of intervention are likely to foster lasting benefits. (see text pages 236-238)

One day, Sammy's mother explained to him that the family would take a vacation in Miami. The next morning, Sammy emerged from his room with belongings spilling out of a suitcase and remarked, "I gotted my bag packed. When are we going to Your-ami?" What do Sammy's errors reveal about his approach to mastering language? (see text page 240)

What can adults do to support children's language development? Provide a list of research-based recommendations. Are those strategies likely to promote academic achievement once the child enters school? Explain. (see text page 241)

SUGGESTED READINGS

Bartsch, K., & Wellman, H. M. (1995). *Children talk about the mind.* New York: Oxford University Press. Considers what children understand about the mind and when this knowledge first emerges. Illustrates children's conceptions by presenting what they say about people and their thoughts.

Fortson, L. R., & Reiff, J. C. (1995). *Early childhood curriculum: Open structures for integrative learning.* Boston: Allyn and Bacon. An excellent, readable practical guide to how to maximize young children's cognitive development while enabling them to develop creatively. Shows how an "open structures" approach can integrate many aspects of knowledge into a central curriculum that promotes intellectual, emotional, and social growth as well as cultural awareness.

Hayman, L. L., Mahon, M. M., & Turner, J. R. (1999). *Health behavior in childhood and adolescence.* Mahwah, NJ: Erlbaum. Presents multidisciplinary research on the interrelationships among behavior, health, and illness in childhood and adolescence within a developmental-contextual framework. Discusses major diseases of childhood and adolescence, chronic childhood conditions, and childhood precursors of adult-onset disease.

Nelson, C. A., & Bloom, F. E. (1997). Child development and neuroscience. *Child Development, 68*, 970–987. Discusses two major advances in developmental neuroscience that have important implications for understanding behavioral development and focuses on bidirectional influences between neurobiological developmental mechanisms and behavior.

Schneider, W., & Pressley, M. (1997). *Memory development between two and twenty* (2nd ed.). Mahwah, NJ: Erlbaum. Presents an overview of the field of memory development; discusses broad topics of general interest such as knowledge, cognitive capacity, and metamemory; and examines controversial issues surrounding memory development, such as children's testimony.

PUZZLE 7.1 TERM REVIEW

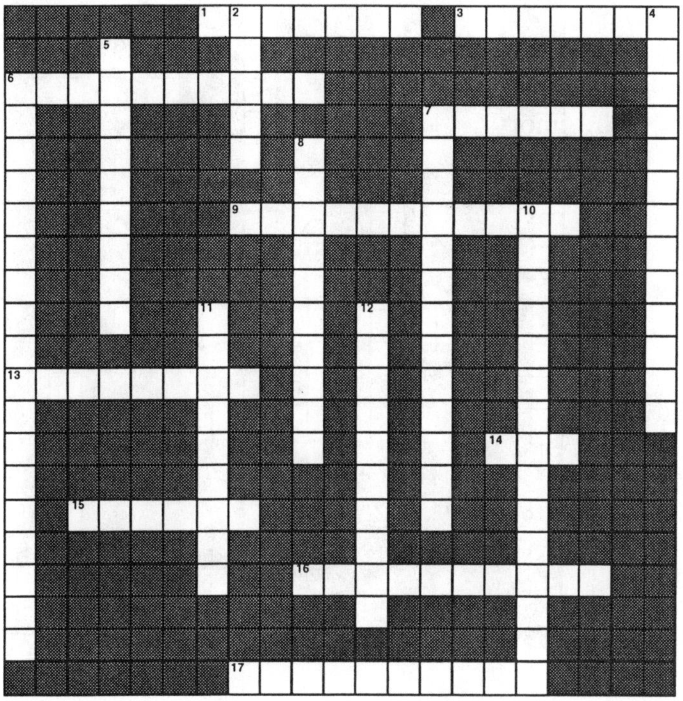

Across

1 General descriptions of what occurs in a particular situation
3 Adult feedback that corrects the form of the child's grammar
6 Concept of order relationships between quantities
7 _____ strategies: deliberate mental activities improve our chances of remembering
9 Math principle stating that the last number in a counting sequence indicates the quantity of items counted
13 _____ preschools: teacher structures the program
14 Therapy that provides sick children with a glucose, salt, and water solution to replace lost fluids (abbr.)
15 _____ hormone: necessary for physical development from birth on
16 Tendency to focus on only one aspect of a situation
17 Practical, social side of language

Down

2 _____-centered preschool: teachers provide a wide variety of activities from which the children select
4 Make-believe play with peers, called _____
5 Gland located at the base of the brain responsible for certain growth hormones
6 _____: application of regular grammatical rules to words that are exceptions
7 Thinking about thought, or _____
8 _____-bound: easily distracted by the appearance of objects
10 Preoperational children treat the initial and final states as completely unrelated events, ignoring the _____.
11 Belief that inanimate objects have life like qualities, called _____ thinking
12 _____: adult feedback that enhances the complexity of the child's statement

PUZZLE 7.2 TERM REVIEW

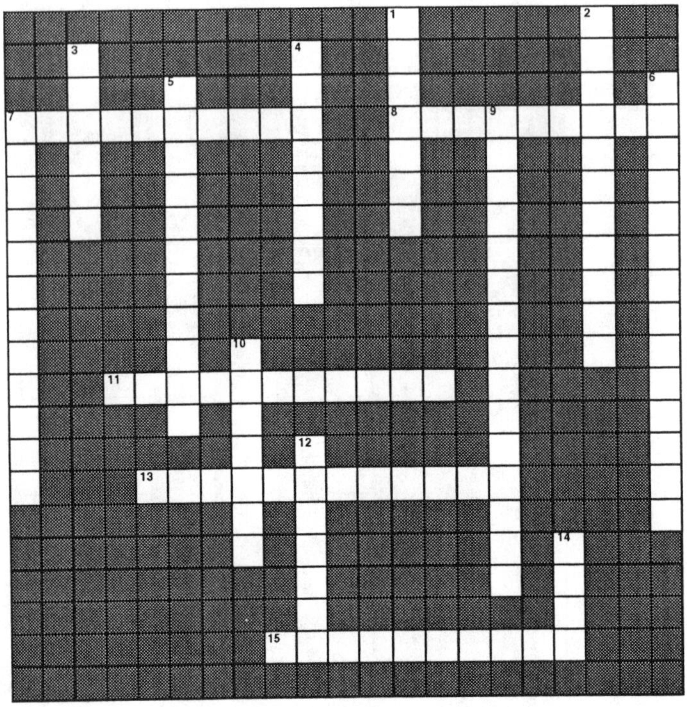

Across

7 Brain structure that aids in balance and control of body movements

8 Brain structure that maintains alertness and consciousness, called the _____ formation

11 Inability to distinguish the viewpoints of others from one's own view

13 Preschoolers have difficulty with _____ classification, or organizing objects into classes and subclasses on the basis of similarities and differences.

15 Project ____ ____: an early intervention program begun by the federal government in 1965 (2 words)

Down

1 ____-stimulating hormone: stimulates the release of thyroxin

2 Assumption that words mark entirely separate categories, called the principle of mutual _____

3 _____ callosum: bundle of fibers that connects the two hemispheres

4 The cerebral hemisphere responsible for skilled motor action

5 Dwarfism caused by emotional deprivation

6 Piaget's second stage

7 Understanding that certain physical characteristics of objects remain the same even when their appearances change

9 Preoperational thought in which children cannot mentally go through a series of steps in a problem and then reverse order

10 ____ growth curve: represents overall changes in body size

12 Self-directed speech used to plan and guide behavior

14 _____-mapping: connecting a new word with a concept after only one encounter

SELF-TEST

1. Which of the following is <u>most</u> helpful in diagnosing growth disorders in early and middle childhood? (209)
   a. universal growth norms
   b. chronological age
   c. skeletal age
   d. developmental quotient

2. Because different body systems have their own unique, carefully timed patterns of maturation, physical development is (209)
   a. an asynchronous process.
   b. a synchronous process.
   c. governed only by genetic factors.
   d. governed only by environmental factors.

3. Language skills increase rapidly in early childhood, and spatial skills develop very gradually over childhood and adolescence. These trends in cognitive development are consistent with (210)
   a. evidence that the two hemispheres of the cortex develop at different rates.
   b. results of brain development studies conducted with lower animal species, such as mice.
   c. what is known about environmental effects.
   d. behaviorist theory.

4. Left-handed persons in the general population are more likely to ____________ than are right-handed persons. (211)
   a. live longer lives
   b. show superior motor coordination by adolescence
   c. have superior verbal and math skills by adolescence
   d. vote in presidential elections

5. During early childhood, the _______________, a bundle of fibers that connects the two hemispheres of the brain, undergoes major changes. (211-212)
   a. corpus callosum
   b. reticular formation
   c. cortex
   d. cerebellum

6. A gland located at the base of the brain that releases hormones affecting physical growth is known as the (213)
   a. sweat gland.
   b. growth gland.
   c. thyroid gland.
   d. pituitary gland.

7. Mario only eats carrots and turkey, and his parents are concerned that he may not be eating enough other foods. What should they do to increase Mario's acceptance of new foods? (214-215)
   a. serve only one new food at each meal, and put a large amount on his plate
   b. continue to expose him to a variety of new foods without insisting that he eat them
   c. tell him that the food has feelings and will cry if he doesn't eat it
   d. serve the new foods first, and give Mario carrots and turkey only after he has eaten all of the new food

8. Which of the following statements is true? (215-216)
   a. Ordinary childhood illnesses have consequences for physical growth among all children.
   b. In well-nourished children ordinary childhood illnesses have no effect on physical growth.
   c. Typically, measles will result in significant weight loss in children.
   d. Among industrialized nations, childhood diseases have increased dramatically during the past 50 years.

9. The leading cause of death among children over 1 year of age is (216)
   a. motor vehicle collisions.
   b. infectious diseases.
   c. drowning.
   d. choking.

10. Boys are more likely to be injury victims than girls because (216)
    a. boys' temperaments are more negative.
    b. girls have better reflexes.
    c. boys are more active and more willing to take risks.
    d. girls are more sensible.

11. While throwing a ball, Mary Grace turns her body only slightly to help increase the distance, and when she catches the ball she usually traps it against her chest. Mary Grace is likely to be a (218)
    a. 2-year-old.
    b. 3-year-old.
    c. 5-year-old.
    d. 6-year-old.

12. The child of 3 to 4 years of age can do all of the following EXCEPT (218)
    a. fasten large buttons.
    b. use a knife to cut food.
    c. use scissors to cut paper.
    d. copy vertical lines and circles.

13. Children's drawings (219-220)
    a. are random scribbling until approximately age 4.
    b. become more realistic as their fine motor skills improve.
    c. become more realistic only as a result of practice.
    d. do not appear to follow a developmental course.

14. Between the ages of 2 and 7, children are at Piaget's _______ stage of development. (221)
    a. formal operational
    b. concrete operational
    c. preoperational
    d. sensorimotor

15. According to Piaget, ________ thinking is illogical, rigid, and unreflective. (223)
    a. egocentric
    b. concrete operational
    c. abstract
    d. fundamental

16. _________, or a child's inability to go through a series of steps and then reverse the steps, is the most important illogical feature of preoperational thought. (224)
    a. centration
    b. irreversibility
    c. reversibility
    d. transformation

17. __________ is/are important in helping preschoolers understand the differences between reality and appearance. (226-227)
    a. Motor development
    b. Cognitive development
    c. Make-believe play
    d. Structured preschool lessons

18. Tanya is a Piagetian first-grade teacher who wants all her students to excel in cognitive development. What will she do to encourage their development? (227)
    a. nothing at all: children will learn successfully in any environment
    b. arrange situations that allow self-paced discovery
    c. reinforce correct answers and ignore incorrect ones
    d. provide explicit verbal training in each subject area

19. Vygotsky viewed preschoolers' private speech as (228)
    a. egocentric and nonsocial.
    b. senseless uttering.
    c. ineffective and noncommunicative.
    d. the foundation for all higher cognitive processes.

20. Vygotsky's approach to education emphasizes (229)
    a. self-initiated discovery learning.
    b. behavior modification techniques.
    c. modeling.
    d. assisted discovery.

21. Like adults, children use ________ to recall familiar, repeated events. (231)
    a. scripts
    b. autobiographical memory
    c. metacognition
    d. hierarchical classification

22. Which of the following is TRUE? (232)
    a. Preschoolers do not understand much about written language until they learn to read and write.
    b. Preschoolers have a great deal of understanding about written language long before they learn to read and write.
    c. Young preschoolers do not search for units of written language.
    d. Children do not begin writing until elementary school.

23. Which of the following is true of child-centered preschools? (236)
    a. The teacher structures the program.
    b. Free play is de-emphasized.
    c. There is a strong focus on letters, number, and colors.
    d. A wide selection of activities is available for children to choose among.

24. Which of the following is NOT one of the four factors that are especially important in determining child care quality? (237)
    a. group size
    b. home-based versus center-based
    c. caregiver/child ratio
    d. caregiver's educational experience

25. Young Colleen says, "I go school, too." Her mother says, "Yes, you are going to school, too." Her mother's response would be classified as (241)
    a. repetition and correction.
    b. correction and reflection.
    c. rejection and restatement.
    d. expansion and recast.

CHAPTER 8

EMOTIONAL AND SOCIAL DEVELOPMENT IN EARLY CHILDHOOD

BRIEF CHAPTER SUMMARY

Erikson's stage of initiative versus guilt offers an overview of the personality changes of early childhood. In early childhood self-concept begins to take shape and children's self-esteem is high, supporting their enthusiasm for mastering new skills. Preschoolers' understanding of emotion, emotional self-regulation, and capacity to experience self-conscious emotions improve, supported by gains in cognition and language as well as warm, sensitive parenting.

During the preschool years, peer interaction increases, cooperative play becomes common, and first friendships are formed, providing important contexts for the development of a wide range of social skills. The development of peer sociability is influenced by cultural variations and parental guidance.

Different aspects of early moral functioning are focused on by three theoretical approaches—psychoanalytic, behaviorist and social learning, and cognitive-developmental. Hostile family atmospheres, poor parenting practices, and heavy television viewing promote childhood aggression, which can spiral into serious antisocial activity.

Gender typing develops rapidly over the preschool years. Heredity contributes to some aspects of gender-typed behavior, but environmental forces, such as parents, siblings, teachers, peers, and television play a powerful role. Neither cognitive-developmental theory nor social learning theory provide a complete account of the development of gender identity. Gender schema theory is a new approach that shows how environmental pressures and children's cognitions combine to affect gender-role development.

Compared to children of authoritarian and permissive parents, children whose parents use an authoritative style are especially well adjusted and socially mature. Warmth, explanations, and reasonable demands for mature behavior account for the effectiveness of the authoritative style. When child maltreatment occurs, it is the combined result of factors within the family, community, and larger culture. Interventions at all of these levels are essential for preventing it.

LEARNING OBJECTIVES

After reading this chapter, you should be able to:

8.1. Describe Erikson's stage of initiative versus guilt, noting major personality changes. (248-249)

8.2. Describe preschoolers' self-concepts and emerging sense of self-esteem. (249-250)

8.3. Describe changes in understanding and expression of emotion during early childhood, noting achievements and limitations. (251-253)

8.4. Trace the development of peer sociability in early childhood, noting cultural variations and the special contribution of sociodramatic play to emotional and social development. (253-255)

8.5. Describe the quality of preschoolers' friendships, and discuss parental influences on early peer relations. (255)

8.6. Compare psychoanalytic, behaviorist and social learning, and cognitive-developmental approaches to moral development, and trace milestones of morality during early childhood along with child-rearing practices that support or undermine them. (256-260)

8.7. Describe the development of aggression in early childhood, discuss family and television as major influences, and cite ways to control aggressive behavior. (260-262)

8.8. Describe preschoolers' gender-stereotyped beliefs and behaviors, and discuss genetic and environmental influences on gender typing. (262-265)

8.9. Describe and evaluate the accuracy of major theories of the emergence of gender identity, and cite ways that adults can reduce gender typing in early childhood. (265-267)

8.10. Describe the impact of child-rearing styles on development, and cite cultural variations in child-rearing beliefs and practices. (267-270)

8.11. List five forms of child maltreatment, and discuss its multiple origins, consequences for development, and prevention strategies . (270-273)

STUDY QUESTIONS

Erikson's Theory: Initiative versus Guilt

1. Erikson regarded ____ as the critical psychological conflict of the preschool years. (248)

2. What did Erikson mean by a sense of initiative? (248)

____

3. Erikson regarded ______________ as a central means through which children find out about their world because it permits them to try out new ______________ with little risk of criticism or failure, and it creates a social organization of children who must ____________________ to achieve common goals. (248)

4. Freud suggested that children develop a _________________, or conscience, in order to manage anxiety, avoid punishment, and maintain parental affection. (249)

5. Erikson regarded an (overly/insufficiently) strict superego as the possible negative outcome of early childhood. (249)

Self-Development

1. Like other aspects of their thinking, preschoolers self-concepts are very __________________, focusing on ____________________ characteristics. (249)

2. The (stronger/weaker) the child's self-definition, the more possessive he or she tends to be about objects, suggesting that early struggles over objects represent an effort to clarify boundaries between __________________ and __________________. (249)

3. True or False: A firmer sense of self underlies the emergence of both conflict and cooperation during the early preschool years. (249)

4. Preschoolers' self-esteem (is/is not) as well-defined as that of older children and adults. (250)

5. Why is preschoolers' very high sense of self-esteem and tendency to underestimate task difficulty adaptive? (250)

____
____

6. List five ways to foster a healthy self-image in preschool children. (250)

A. ____
B. ____
C. ____
D. ____
E. ____

Emotional Development

1. Young children's vocabulary for talking about emotion expands rapidly. To what three aspects of language do they often refer? (251)

A. ______________________________ B. ______________________________
C. ______________________________

2. Four- and five-year olds (do/do not) realize that thinking and feeling are interconnected. (251)

3. Young children also use emotional language to __________________________ a companion's behavior. (251)

4. However, preschoolers have difficulty making sense of what is going on when there are ______________ clues about how a person is feeling. (251)

5. By age 3 to 4, children verbalize a variety of strategies for controlling emotional arousal. List three such strategies. (251)

A. ______________________________ B. ______________________________
C. ______________________________

6. Along with language, ______________________ and ______________________ also affect children's development of emotional self-regulation. (251)

7. How can parents support the development of emotional self-regulation in early childhood, including the management of fear? (251-252)

____
____
____

8. List four indicators that a child's fear has reached the level of a phobia. (252)

A. ______________________ B. ______________________

C. ______________________ D. ______________________

9. List three ways in which preschoolers' experience of self-conscious emotions differs from that of older children and adults. (252)

A. ______________________

B. ______________________

C. ______________________

10. Compared to toddlers, preschoolers rely more on ______________ to console others and express empathy more often. (252)

11. Empathy is related to both ______________ and ______________. (252-253)

Peer Relations

1. Briefly describe Parten's three-step sequence of social development. (253-254)

A. ______________________

B. ______________________

C. ______________________

2. Parten's play forms (do/do not) emerge in order and (do/do not) follow a developmental sequence in which earlier ones are replaced with older ones. (254)

3. It is the ______________ rather than the ______________ of solitary and parallel play that changes in early childhood. Research indicates that within each of Parten's play types, older children display more ______________ than do younger children. (254)

4. List three types of nonsocial activity that are cause for concern during the preschool years. (254)

A. ______________________ B. ______________________

C. ______________________

5. List three aspects of sociodramatic play that contribute to cognitive and social development. (254)

A. ______________________

B. ______________________

C. ______________________

6. True or False: Social interaction is essentially the same across cultures, regardless of differing cultural beliefs. (254-255)

7. Preschoolers know that a friend is someone who ______________ and with whom you ______________, but their friendships do not have a long-term, enduring quality based on ______________. (255)

8. Preschool friends give and receive more ____________________ and are more ____________________ expressive to each other than to others. (255)

9. List two ways parents can influence early peer relations. (255)
A.____________________
B.____________________

Foundations of Morality

1. Briefly cite examples of the young child's developing sense of morality. (256)
____
____
____

2. All theories of moral development recognize that conscience begins to take shape in early childhood and most agree that the child's morality is first ____________________ controlled by adults and gradually becomes regulated by ____________________. (256)

3. Match each of the following major theories of moral development with the aspect of moral functioning it emphasizes: (256)

____	Emotional side of conscience	1. Behaviorism
____	Ability to reason about justice and fairness	2. Psychoanalytic theory
____	Moral behavior	3. Cognitive-developmental theory

4. Children whose parents use threats, commands, or physical force usually feel (intense/little) guilt after harming others. (256)

5. List three reasons why induction promotes conscience formation. (256-257)
A. ____________________
B. ____________________
C. ____________________

6. Give examples of maternal discipline that predict conscience development in inhibited children and impulsive children. (257)
Inhibited:____________________
Impulsive:____________________

7. True or False: Guilt (is/is not) an important factor in the development of early conscience. (257)

8. Why is operant conditioning not a sufficient explanation for children's acquisition of many moral responses? (258)
____

9. List three characteristics of models that affect children's willingness to imitate them. (258)
A. ____________________
B. ____________________
C. ____________________

10. True or False: Models exert their most powerful effect on prosocial development during the preschool years. (258)

11. List three undesirable side effects of harsh punishment? (258)
A. ______________________________
B. ______________________________
C. ______________________________

12. List two alternatives to harsh punishment. (258)
A. ______________________ B. ______________________

13. List three ways that parents can increase the effectiveness of punishment when they do decide to use it. (258-259)
A. ______________________ B. ______________________
C. ______________________

14. Parenting practices that ______________________ good conduct are the most effective forms of discipline. (259)

15. What feature do psychoanalytic and behaviorist approaches to morality have in common, and how does the cognitive-developmental perspective differ in this respect? (259-260)

____
____
____

16. List two examples of preschoolers' moral understanding. (260)
A.______________________________
____
B.______________________________
____

17. According to cognitive-developmental theorists, preschoolers learn to distinguish intentional harm from accidents and moral transgressions from violations of social conventions in which of the following ways? (260)
A. Through direct teaching, modeling, and reinforcement
B. By actively making sense of their observations

18. List three ways of handling rule violations and moral discussions which are associated with children's advanced moral thinking. (260)
A. ______________________________
B. ______________________________
C. ______________________________

19. The most common form of aggression in the preschool years is ______________ aggression. The other type of aggression is ______________________ aggression and can be divided into two varieties: ______________________ aggression and ______________________ aggression. (260)

20. Instrumental aggression (increases/decreases) with age and hostile aggression (increases/decreases) between 4 and 7. (260)

21. Cite evidence indicating that the trend for boys to be more overtly aggressive than girls is due to both biological and environmental factors. (261)
Biological: ____
____
Environmental: ____
____
____

22. Preschool and school-age girls are most likely to express their hostility through ______________________ aggression. (261)

23. Describe the pattern and negative consequences of a hostile family atmosphere. (261)
____
____
____
____
____

24. List two reasons that boys are more likely than girls to become involved in family interactions that promote aggressive behavior. (261)
A. ____
B. ____

25. Briefly describe the ways violence is often depicted on TV. (261)
____
____

26. Young children are especially likely to be influenced by television because they do not ____________________ a great deal of what they see on TV and because they find it difficult to separate ________________ from ______________ television content. (261)

27. TV violence is related to increased hostile behavior in highly _______________ children and hardens children to violence, making them more likely to ____________________________________. (262)

28. The ______________________________________ has made the federal government reluctant to place limits on the content of television in support of children's needs. However, what steps has the federal government taken to regulate the negative impact of television? (263)
____
____
____

29. List ways in which parents can protect children from the impact of harmful TV. (263)
A. ____
B. ____
C. ____
D. ____

30. Cite several ways parents and teachers can help control children's aggression. (262)
____
____
____

Gender Typing

1. Preschoolers' rigid gender-stereotyped beliefs are a joint product of ________________________________ in the environment and children's ________________________________. (264)

2. What two aspects of children's gender-typed behavior are believed to be partly influenced by hormones? (264)
A. ______________________________ B. ______________________________

3. Describe ways in which parents encourage gender-stereotyped beliefs and behavior in their children. (264-265)

____

____

____

____

____

4. As at home, girls get more encouragement to participate in ______________ activities at preschool. (265)

5. True or False: Preschool peers, especially boys, reinforce gender-stereotyped behavior and criticize "gender-inappropriate" behaviors. (265)

6. In cartoon programs, male characters are more likely to be portrayed as ________________________, whereas female characters are portrayed as ________________________. (265)

7. Masculine and androgynous individuals have higher _______________ than do feminine individuals. (266)

8. According to social learning theory, (behaviors/self-perceptions) come before (behaviors/self-perceptions) in the formation of gender-role identity. Cognitive-developmental theory views gender-role formation as occurring in the (same/opposite) direction. (266)

9. Indicate when gender constancy develops in most children. (266)

____

10. Preschoolers' poor performance on gender constancy tasks is not only the result of cognitive immaturity, but also of a lack of opportunity to learn about _______________ differences between the sexes. (266)

11. "Gender-appropriate" behavior occurs so early in the preschool years that ________________________________ must account for its initial appearance. (266)

12. Describe *gender schema theory.* (266)

____

____

____

____

13. In order to support the goal of the "equal roles family model," Sweden has made ______________________________ available outside the home and mandated that parents of children under the age of 8 could ________________________________ with a reduction in pay but not in benefits. (268)

14. As a result of these and other equality-supporting policies, gender roles in Sweden are _________________________, child rearing is more _____________, and adolescents are (more/less) likely to value "masculine" over "feminine" roles than elsewhere in the world. (268)

15. List ways in which adults can help children avoid developing rigid gender-typed beliefs and behaviors. (267)

____
____
____
____
____

Child Rearing and Emotional and Social Development

1. Name and describe the two dimensions of parenting that emerged from Baumrind's observations of parents interacting with their preschoolers. (269)

A.____
____

B.____
____

2. Match each of the following parenting styles with its appropriate description. (269)

_____	High demandingness, low responsiveness	1. Authoritative
_____	High demandingness, high responsiveness	2. Authoritarian
_____	Low demandingness, low responsiveness	3. Permissive
_____	Low demandingness, high responsiveness	4. Uninvolved

3. Briefly describe child outcomes associated with each style of parenting. (269)

Authoritative: ____
____

Authoritarian: ____
____

Permissive: ____
____

4. Cite three reasons that authoritative parenting is especially effective. (269)

A.____
____

B.____
____

C.____
____

5. Describe how the parenting practices of the following cultural groups often differ from those of Caucasian Americans. (270)
Chinese: ____
Hispanic: ____
African-American: ____

6. At its extreme, uninvolved parenting is a form of child maltreatment called ________________. (270)

7. A total of ________________ cases of child maltreatment were reported to juvenile authorities in 1997, an increase of ______ percent over the previous decade. Reported cases of child maltreatment (overestimate/underestimate) the true number. (270)

8. List five forms of child maltreatment. (270)
A. ______________________________ B. ______________________________
C. ______________________________ D. ______________________________
E. ______________________________

9. True or False: The most frequent form of child abuse is most likely psychological. (271)

10. True or False: Child abuse is rooted in a single abusive personality type. (271)

11. List some characteristics of children and parents that increase the likelihood that abuse will occur. (271)
Children: ____
____
Parents: ____
____
____

12. List family factors that combine with child and parent characteristics to prompt child maltreatment. (271)
____
____
____

13. Most abusive parents are isolated from supportive ties to the community because they have learned to ______________________________ and because they are likely to live in neighborhoods that provide ______________________ between family and community. (272)

14. The ________________, ________________, and ________________ of a society have profound effects on the chances of child maltreatment. In countries where ______________________ is not accepted, child abuse is rare. (272)

15. Describe the developmental consequences of child maltreatment. (272)
____
____
____

16. List three approaches that can help prevent child maltreatment. (272-273)
A. ____________________
B. ____________________
C. ____________________

ASK YOURSELF . . .

Reread the description of Sammy and Mark's argument at the beginning of this chapter. On the basis of what you know about self-development, why was it a good idea for Leslie to resolve the dispute by providing an extra set of beanbags so that both boys could play at once? (see text pages 248-249)

____
____
____

Four-year-old Tia had just gotten her face painted at a carnival. Soon the heat of the afternoon caused her balloon to pop. When Tia started to cry, her mother said, "Oh, Tia, balloons aren't such a good idea when it's hot outside. We'll get another on a cooler day. If you cry, you'll mess up your beautiful face painting." What emotional capacity is Tia's mother trying to promote, and why is her intervention likely to help Tia? (see text page 251)

____
____
____

How does emotional self-regulation contribute to the development of empathy and sympathy? Why are these emotional capacities vital for positive peer relations? What can parents do in early childhood to promote favorable peer ties? (see text pages 252-255)

____
____
____
____

Alice and Wayne want their two young children to develop a strong, internalized conscience and to become generous, caring individuals. List as many parenting practices as you can that would foster these goals. (see text pages 256-258)

____
____
____
____

Nanette told her 3-year-old son Darren not to go into the front yard without asking, since the house faces a very busy street. Darren disobeyed several times, and now Nanette thinks it's time to punish him. How would you recommend that Nanette discipline Darren, and why? (see text page 258-259)

____
____
____
____

Suzanne has a difficult temperament, and her parents respond to her angry outbursts with harsh, inconsistent discipline. Explain why Suzanne is at risk for long-term difficulties in conscience development and peer relations. (see text pages 257-259, 261)

Geraldine cut her 3-year-old daughter Fern's hair very short for the summer. When Fern looked in the mirror, she said, "I don't wanna be a boy," and began to cry. Why is Fern upset about her short hairstyle, and what can Geraldine do to help? (see text pages 265-266)

When 4-year-old Roger was in the hospital, he was cared for by a male nurse named Jared. After Roger recovered, he told his friends about Dr. Jared. Using gender schema theory, explain why Roger remembered Jared as a doctor, not a nurse. Why are gender schemas likely to persist, affecting children's thinking and behavior for years to come? (see text page 266)

Earlier in this chapter, we discussed induction as an especially effective form of discipline. Of Baumrind's three child-rearing styles, which is most likely to be associated with use of induction, and why? (see text pages 256-257, 269)

Chandra heard a news report that 10 severely abused and neglected children, living in squalor in an inner-city tenement, were discovered by Chicago police. Chandra thought, "What could possibly lead parents to mistreat their children so badly?" How would you answer Chandra's question? Why are these children likely to suffer lasting impairments to development? (see text pages 271-273)

SUGGESTED READINGS

Brannon, L. (1998). *Gender: Psychological perspectives*. Boston: Allyn and Bacon.
Summarizes the literature on gender differences and includes personal narrative accounts of gender-relevant real-life experiences. Topics include the brain, health and fitness, gender stereotypes, discrimination, sexual harassment, and emotion.

Dunn, Judy (1988). *The beginnings of social understanding.* Cambridge, MA: Harvard University Press. A keen observer of children's everyday social behavior, Dunn explores young children's ability to recognize feelings, predict the behavior of others, and grasp the rules of social interaction. She shows that 3- to 5-year-olds' social understanding is far more advanced than previously believed.

Gullotta, T. P., Adams, G. R., & Montemayor, R. (Eds.). (1998). *Delinquent violent youth.* Thousand Oaks, CA: Sage. Provides an overview of crime among both urban and rural youths. Issues include how various social factors influence delinquency, treatment for violent behavior, and social policies that prevent crime.

Maccoby, E. E. (1994). The role of parents in the socialization of children: An historical overview. In R. D. Parke, P. A. Ornstein, J. J. Rieser, & C. Zahn-Waxler (Eds.), *A century of developmental psychology.* Washington, DC: American Psychological Association. Discusses the past, present, and future of research on the role of the family in children's socialization.

Rochat, P. (1999). *Early social cognition.* Mahwah, NJ: Erlbaum. Edited volume discussing current conceptualizations and research on the developmental origins of social cognition.

PUZZLE 8.1 TERM REVIEW

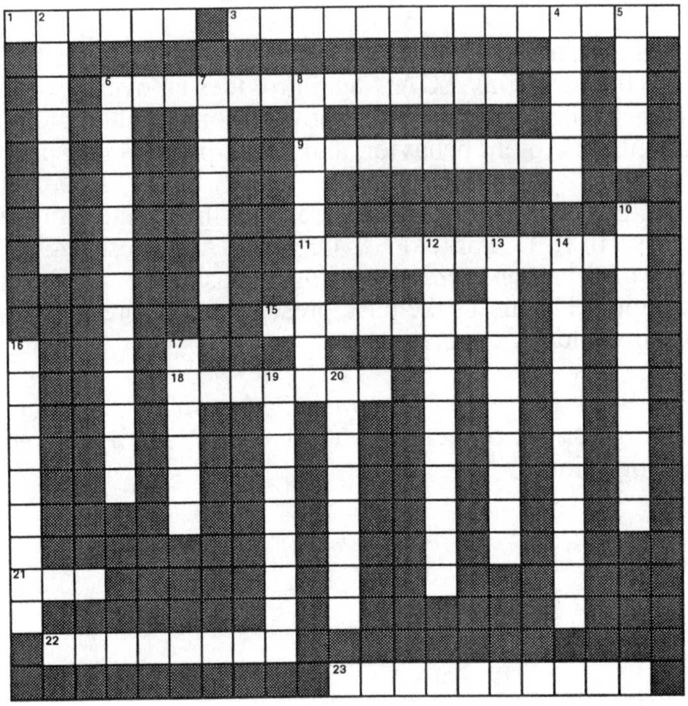

Across

1 Self-____: the judgments we make about our own worth and the feelings associated with those judgments
3 Process of forming a superego by taking the same-sex parent's characteristics into one's personality
6 Parenting style that is both demanding and responsive
9 Discipline communicating the effects of the child's misbehavior on others
11 Play involving separate activities but exchanging of toys and comments
15 Mild punishment involving removal of the child from the setting until ready to behave appropriately (2 words)
18 Self-____: attributes, abilities, and attitudes that an individual believes defines who he/she is
21 _____-social activity includes unoccupied, onlooker behavior and solitary play.
22 Gender _____: an image of oneself as relatively masculine or feminine
23 Aggression which damages another's peer relationships

Down

2 Feelings of concern or sorrow for another's plight
4 Gender _____: process of developing gender roles
5 Aggression which harms others through physical injury
6 Parenting style that is demanding but unresponsive
7 Aggression intended to harm another
8 Positive outcome of Erikson's conflict of early childhood
10 Parenting style that is responsive but undemanding
12 Play with others directed at a common goal
13 Behavior benefiting another without expected reward
14 Aggression aimed at obtaining an object, privilege, or space with no intent to harm
16 A type of gender-role identity high in both masculine and feminine characteristics
17 Gender _____ theory: combines social learning and cognitive-developmental features
19 Gender _____: understanding that sex remains the same even if outward appearance changes
20 Play that occurs near other children with similar materials, but no interaction

SELF-TEST

1. Make-believe play (248-249)
   a. is a cultural phenomenon seen only in the United States.
   b. provides children with insights into the link between self and society.
   c. is of little help in resolving the initiative versus guilt conflict.
   d. is very fantasy-oriented, having little or no connection with the real world.

2. Erikson saw the negative outcome of the initiative versus guilt stage as closely related to Freud's notion of (249)
   a. emotional self-regulation.
   b. hostile aggression.
   c. an overly confident self-image.
   d. an overly strict superego.

3. Preschoolers' self-concepts are based primarily on observable characteristics and are therefore (249)
   a. internal.
   b. concrete.
   c. made up of stable personality traits.
   d. abstract.

4. Preschoolers tend to (250)
   a. overestimate the difficulty of tasks.
   b. appraise the difficulty of tasks accurately.
   c. underestimate their abilities.
   d. overestimate their abilities.

5. Children have the capacity to correctly judge the causes of many basic emotions by age (251)
   a. 2 to 3.
   b. 3 to 4.
   c. 4 to 5.
   d. 5 to 6.

6. Three- to 4-year-olds who have trouble controlling their anger and hostility (251)
   a. probably have parents who model ineffective regulation of their own emotions.
   b. have probably picked up their antisocial behavior from peers.
   c. enjoy hurting others.
   d. are too young to have acquired strategies for emotional self-regulation.

7. Self-conscious emotions (252)
   a. are labeled precisely by preschoolers.
   b. are not dependent on the presence of an audience for preschoolers.
   c. involve injury to or enhancement of the sense of self.
   d. are not linked to self-evaluation in the preschool years.

8. According to recent research, ______________ preschoolers rarely show any signs of empathy. (253)
   a. nonabused
   b. highly socialized
   c. physically abused
   d. overly active

9. According to Parten, the first form of true social interaction to develop during early childhood is (253)
   a. cooperative play.
   b. associative play.
   c. parallel play.
   d. onlooker behavior.

10. Preschoolers who are friends (255)
   a. do not have any explanation for why they are friends.
   b. know that they trust one another.
   c. are more emotionally expressive with each other than with non-friends.
   d. describe their relationship in abstract terms.

11. Identification and guilt as motivators of good conduct are stressed by ____________ theory. (256-257)
   a. psychoanalytic
   b. behaviorist
   c. ecological
   d. cognitive-developmental

12. Researchers have found that observing a model performing prosocial behaviors (258)
   a. is not effective in encouraging prosocial behaviors in preschoolers.
   b. has no long-term effects on children's prosocial behavior.
   c. exerts the greatest influence on prosocial behavior during the preschool years.
   d. exerts the greatest influence on prosocial behavior during middle childhood.

13. The most effective form of discipline (259)
   a. is withdrawal of privileges.
   b. is time out.
   c. encourages good conduct.
   d. is punishment.

14. The _________________ view of morality regards children as active thinkers about social rules. (260)
   a. psychoanalytic
   b. cognitive-developmental
   c. behaviorism and social learning
   d. ecological

15. For most preschoolers, instrumental aggression ________ with age and hostile aggression ________. (260)
   a. declines; increases
   b. increases; declines
   c. declines; declines
   d. increases; increases

16. Because they expect others to react with anger and physical force, children from conflict-ridden families (261)
   a. make many unprovoked attacks.
   b. see hostile intent where it does not exist.
   c. neither a nor b is often the case
   d. both a and b are often the case

17. TV violence (261-262)
   a. increases children's sensitivity to violent acts around them, making them less tolerant of them.
   b. leads children to view aggressive acts as normal and acceptable means for solving problems.
   c. does not produce aggression, but stimulates children's innate desire to be aggressive.
   d. is not as common today in children's shows.

18. According to Eleanor Maccoby, ______________ have an important effect on gender-typed play behaviors. (264)
   a. learning styles
   b. beliefs
   c. parental occupations
   d. hormones

19. In preschool (265)
   a. girls and boys are equally encouraged to participate in adult-structured activities.
   b. boys are encouraged to participate in more adult-structured activities.
   c. girls are encouraged to participate in more adult-structured activities.
   d. neither boys nor girls are encouraged to participate in adult-structured activities.

20. Children and adults who exhibit either masculine or androgynous characteristics (266)
   a. are less well-adjusted than "feminine" individuals.
   b. are higher in self-esteem than "feminine" individuals.
   c. are very rare.
   d. do not possess any feminine characteristics.

21. At the end of the preschool years, when children understand that their sex is a permanent characteristic of the self, they have developed (266)
   a. gender constancy.
   b. sexual stability.
   c. gender identity.
   d. gender neutrality.

22. According to gender schema theory, information that is _________________ will be remembered. (266)
   a. redundant with the schema
   b. consistent with the schema
   c. moderately discrepant from the schema
   d. contradictory to the schema

23. ___________________ are demanding and unresponsive and have children who __________________________. (269)
   a. Authoritarian parents; are anxious and unhappy
   b. Permissive parents; have difficulty controlling impulses
   c. Uninvolved parents; are severely maladjusted
   d. Authoritative parents; are happy and self-confident

24. The largest number of child sexual abuse cases are identified in (271)
   a. toddlerhood.
   b. early childhood.
   c. middle childhood.
   d. adolescence.

25. Which of the following is strongly associated with all forms of child abuse? (271)
   a. the abuser having been abused as a child
   b. unmanageable stress in the abuser
   c. a personality disturbance on the part of the abuser
   d. a sick or premature baby

CHAPTER 9

PHYSICAL AND COGNITIVE DEVELOPMENT IN MIDDLE CHILDHOOD

BRIEF CHAPTER SUMMARY

The slow gains in body growth that took place during the preschool years continue in middle childhood. Bones of the body lengthen and broaden, and primary teeth are replaced with permanent teeth. Although many school-age children are at their healthiest, health problems do occur. While they can learn a wide range of health information, school-age children do not find it relevant enough to impact their daily behavior. Improvements in gross motor skills occur as a result of gains in flexibility, balance, agility, force, and reaction time, and fine motor coordination increases. Rule-oriented games become common, and physical education helps ensure that all children have access to regular exercise and play.

Thought becomes more logical, flexible, and organized during Piaget's concrete operational stage, but children cannot yet think abstractly. Research findings raise questions about whether mastery of Piagetian tasks emerges spontaneously, about his assumption of an abrupt, stagewise transition to logical thought, and about the impact of specific cultural and school practices.

During middle childhood, attention becomes more adaptable and planful, and memory strategies improve. Metacognition moves from a passive to an active view of the mind. Still, school-age children have difficulty regulating their progress toward goals. Reading and mathematics instruction that combine conceptual understanding with basic skills training may be most effective.

Intelligence tests for children measure overall IQ and several separate intellectual factors. Sternberg's triarchic theory extends our understanding of the determinants of intelligence, and both genetic and environmental factors contribute to individual differences in IQ. The IQ scores of low-SES minority children often underestimate their true abilities. Gardner's theory of multiple intelligences provides yet another view of how information-processing skills underlie intelligent behavior.

Although less dramatic, language development continues during the school years. Vocabulary increases rapidly and pragmatic skills are refined. Bilingual children are advanced in cognitive and language development, although controversy has arisen over the best way to approach bilingual education.

Class size, teacher's educational philosophy and interaction with pupils, and grouping practices have an important impact on learning. Teaching children with learning disabilities or special intellectual strengths introduces unique challenges. American pupils have not fared well in international comparisons of academic achievement. Efforts are underway to improve American education.

LEARNING OBJECTIVES

After reading this chapter, you should be able to:

9.1. Describe changes in body size, proportions, and skeletal maturity during middle childhood. (280-281)

9.2. Identify common health problems in middle childhood, discuss their causes and consequences, and cite ways to alleviate them. (281-286)

9.3. Cite major milestones of gross and fine motor development in middle childhood, noting sex differences. (286-288)

9.4. Describe qualities of children's play during middle childhood, along with consequences for cognitive and social development. (288-289)

9.5. Discuss the importance of high-quality physical education during the school years. (290)

9.6. Describe the major characteristics of concrete operational thought. (290-291)

9.7. Discuss recent research on concrete operational thought and its implications for the accuracy of Piaget's concrete operational stage. (292-293)

9.8. Cite two basic changes in information processing, and describe changes in attention during middle childhood. (293-294)

9.9. Describe the development of memory strategies in middle childhood, and discuss the role of knowledge and culture in memory performance. (295-296)

9.10. Describe the school-age child's theory of mind and capacity to engage in cognitive self-regulation. (296-297)

9.11. Discuss current controversies in teaching reading and mathematics to elementary school children. (297-298)

9.12. Describe major approaches to defining and measuring intelligence, including Sternberg's triarchic theory and Gardner's theory of multiple intelligences. (299-302)

9.13. Describe evidence indicating that both heredity and environment contribute to IQ, and discuss cultural influences on mental test scores of ethnic minority children. (302-304)

9.14. Describe changes in vocabulary, grammar, and pragmatics during middle childhood, and discuss the advantages of bilingualism. (304-306)

9.15. Describe the impact of class size and major educational philosophies on children's motivation and academic achievement. (306-308)

9.16. Discuss the role of teacher-pupil interaction and grouping practices in academic achievement. (308-309)

9.17. Discuss the conditions under which placement of children with mild mental retardation and learning disabilities in regular classrooms is successful. (309-310)

9.18. Describe the characteristics of gifted children and current efforts to meet their educational needs. (310-311)

9.19. Compare the American cultural climate for academic achievement with that of Asian nations. (311-313)

STUDY QUESTIONS

Physical Development

Body Growth

1. During middle childhood, children continue to add about __________ inches in height and ____ pounds in weight each year. However, rather than occurring steadily, growth occurs in ________________. (280)

2. Girls are slightly shorter and lighter than boys at ages ________________. This trend ________________ by age 9. (280)

3. What portion of the body grows fastest during middle childhood? _________. (280)

4. List two factors that account for unusual flexibility of movement in middle childhood. (280-281)
A. ________________________________ B. ________________________________

5. Between the ages of _____ and _____, all primary teeth are replaced by permanent teeth. The first teeth to go are the ______________________. (281)

Common Health Problems

1. List two factors that lead many children from advantaged homes to be at their healthiest in middle childhood. (281)
A. ________________________________ B. ________________________________

2. ______________ continues to be a powerful predictor of ill health. (282)

3. Cite evidence indicating that nearsightedness is not only hereditary, but also related to environment. (282)
____
____

4. About ________ percent of the school-age population and _________ percent of low-SES children develop some hearing loss due to repeated ear infections. (282)

5. List five ways in which the prolonged effects of malnutrition are apparent by middle childhood. (282)
A. ________________________________ B. ________________________________
C. ________________________________ D. ________________________________
E. ________________________________

6. About _____ percent of American children are obese. (282)

7. (Low-SES/middle-SES) children are more likely to be overweight. Cite three factors that are responsible for this trend. (282-283)
A. ____
B. ____
C. ____

8. Research has revealed a link between growth stunting due to ____________________ and childhood ______________________. List two physiological changes that researchers believe are involved. (283)

A.____

B.____

9. Describe parenting practices that contribute to obesity along with their consequences for eating behaviors. (283-284)

____

____

____

10. Next to already existing obesity, time spent ___________________________ is the best predictor of future obesity among school-age children. (284)

11. Describe the effects of childhood obesity on emotional and social development. (284)

____

____

____

12. Why is childhood obesity difficult to treat? (284)

____

13. List two characteristics of the most effective programs for treating childhood obesity. (284)

A. ______________________________ B. ____________________________________

14. Name the two most frequent causes of enuresis. (285)

A. ____

B. ____

15. The most effective treatment for enuresis is a urine alarm that works according to _____________________________ principles. (285)

16. The most frequent cause of school absence and childhood hospitalization is _________________________. (285)

17. Describe the characteristics of children at greatest risk for asthma. (285)

____

____

____

18. True or False: During middle childhood, the frequency of unintentional injuries increases, with boys continuing to show a higher rate than girls. (285)

19. Describe child-rearing techniques that tend to be used by parents of children who take the most risks. (286)

____

____

Motor Development and Play

1. Improvements in motor skills over middle childhood reflect gains in what four capacities? (286)
A. ______________________ B. ______________________
C. ______________________ D. ______________________

2. Body growth, as well as more efficient ______________________, contributes to improved motor performance in children. (286)

3. At age 6, writing tends to be quite large because children use the ______________________ to make strokes rather than just the wrist and fingers. Fine motor skills gradually improve and prepare children to master cursive writing by __________ grade. (286)

4. In what ways do children's drawings improve over the school years? (288)
____
____

5. Girls remain ahead of boys in the fine motor area and gross motor skills that depend on ______________________ and ______________________. (288)

6. True or False: School-age boys' genetic advantage in muscle mass is great enough to account for their superiority in most gross motor skills. (288)

7. Name two special measures that can help raise girls' sports-related confidence. (288)
A. ______________________ B. ______________________

8. What cognitive capacity permits the transition to rule-oriented games in middle childhood? (289)
____

9. Briefly describe how child-organized games contribute to development. (289)
____
____
____

10. List two factors that have contributed to the recent decline in child-organized games. (289)
A. ______________________ B. ______________________

11. List four arguments against and four arguments in favor of adult-organized sports for children. (289)
Against:
A. ______________________
B. ______________________
C. ______________________
D. ______________________
In favor:
A. ______________________
B. ______________________
C. ______________________
D. ______________________

12. Only _____ of American elementary school pupils have a daily physical education class. The average school-age child gets only _____ minutes of physical education per week. (290)

13. Only _____ of 10- to 12-year-old boys and ______ of 10- to 12-year-old girls meet basic fitness standards for children their age. (290)

14. What kinds of activities should physical education classes emphasize to help the largest number of children develop active and healthy life styles? (290)

____

____

Cognitive Development

Piaget's Theory: The Concrete Operational Stage

1. During Piaget's concrete operational stage, thought is more ___________________, ___________________, and ___________________ than it was during early childhood. (291)

2. What two characteristics of thinking does conservation illustrate? (291)
A. _________________________________ B. _________________________________

3. What characteristic of concrete operational thought is illustrated by school-age children's elaborate sorting of classes and collections? (291)

____

4. Distinguish the performance of a concrete operational child from that of a preoperational child on seriation and transitive inference problems. (291)
A. Seriation: ____

____

B. Transitive inference:____

____

5. When asked to name an object to the left or to the right of a person in a different orientation, 5- and 6-year-olds answer (correctly/incorrectly) and 7- and 8-year-olds answer (correctly/incorrectly). Around 8 to 10 years, children can give (clear/vague) directions for how to get from one place to another. (291)

6. Describe the major limitation of concrete operational thought. (292)

____

7. As the horizontal décalage illustrates, school-age children (do/do not) grasp conservation as a general principle and then apply it universally. (292)

8. Conservation in tribal and village societies is often ________________________. What does this suggest children must do to master conservation and other Piagetian concepts? (292)

____

9. The development of operational thinking may best be understood in terms of gains in ____________________________________ capacity rather than a sudden shift to a new stage. (293)

10. A growing number of researchers believe that both ____________________ and ________________________ change may characterize cognitive development in middle childhood. (293)

Information Processing

1. Brain development contributes to an increase in ______________________ and gains in ____________________________ to facilitate more effective information processing. (293)

2. List the three ways in which attention improves in middle childhood. (293-294)
A. ________________________________ B. ________________________________
C. ________________________________

3. Describe the characteristics of children with attention-deficit hyperactivity disorder (ADHD). (294)
____
____
____

4. ___________________ plays a major role in ADHD; however, ________________ factors are also associated with ADHD. In addition, the family stress to which the child's ADHD contributes may ________________ the child's preexisting problems. (294)

5. The most common treatment for ADHD is ______________________________.
List two additional interventions that help ADHD children. (295)
A. ____
B. ____

6. List the areas in which adults with ADHD may need help. (295)
____
____

7. List three memory strategies that develop in middle childhood in the order in which they typically appear. (295)
A. ________________________________ B. ________________________________
C. ________________________________

8. Because organization and elaboration combine items into ______________________ __________________, they permit children to hold onto much more information at once. (296)

9. Explain how knowledge and memory strategies mutually support one another. (296)
____
____
____

10. According to cross-cultural research, the development of memory strategies is not just a product of improved information processing, but also of _______________ demands and ____________________ circumstances. (296)

11. Provide some examples of the school-age child's understanding of the impact of psychological factors on cognitive performance. (297)

____
____
____
____

12. True or False: School-age children's theory of mind does not yet take account of the combined effects of characteristics of the learner, mental processes, and situational factors on cognitive performance. (297)

13. By secondary school, children's self-regulation is a strong predictor of ______________________. (297)

14. List three ways in which parents and teachers can facilitate children's self-regulatory skills. (297)

A. ____
B. ____
C. ____

15. List the diverse information-processing skills that contribute to reading. (297)

____
____
____

16. Summarize the two sides of the "great debate" over how to teach beginning reading. (297-298)

A.____
____
____

B.____
____
____

17. True or False: A balance between the two approaches to reading instruction may be most effective. (298)

18. Mathematics, as taught in many classrooms, (does/does not) make good use of children's basic grasp of number concepts. (298)

19. The most effective form of mathematics instruction is probably a blend of drill in __________________ skills and instruction aimed at building upon ______________ ______________________. (298)

20. List four factors that support the acquisition of mathematical knowledge in Asian countries. (298)

A. ____
B. ____
C. ____
D. ____

Individual Differences in Mental Development

1. In factor analysis, items that are strongly ________________ are assumed to ______________________________________, and are therefore designated as a single factor. (299)

2. Which of the two intelligence tests discussed in your text measures general intelligence and four intellectual factors: verbal reasoning, quantitative reasoning, spatial reasoning, and short-term memory? (299)

____

3. Which of the two intelligence tests discussed in your text measures general intelligence and two intellectual factors: verbal and performance? (299)

____

4. What is meant by *componential analyses* of children's IQ scores? (299-300)

____
____
____

5. Sternberg's triarchic theory of intelligence expanded the componential approach by regarding intelligence as a product of both ________________ and ________________ forces. (300-301)

6. Match each of the following subtheories of Sternberg's theory with the appropriate descriptions. (300-301)

_____	Ability to process information in novel situations	1. Componential Subtheory
_____	Underlying skills: strategy application, knowledge acquisition, metacognition, and self-regulation	2. Experiential Subtheory
_____	Ability to adapt skills, shape the environment, and select contexts consistent with personal goals	3. Contextual Subtheory

7. List the eight intelligences included in Gardner's theory of multiple intelligences. (301)

A. ______________________ B. ______________________
C. ______________________ D. ______________________
E. ______________________ F. ______________________
G. ______________________ H. ______________________

8. True or False: Because it is not firmly grounded in research, Gardner's theory has not been especially helpful in efforts to understand and nurture children's special talents. (302)

9. The average IQ difference between American black and white children is _____ points. The average IQ difference between middle- and low-SES children is _____ points. (302)

10. Evidence from kinship studies suggests that about _________ the differences among children in IQ can be traced to their genetic makeup. (302)

11. Children of low-IQ biological mothers adopted into affluent homes score (above/below) average in IQ. They (did/did not) perform as well as children of high-IQ biological mothers adopted into affluent homes. (302-303)

12. Black children adopted into affluent white homes score (high/low/average) on intelligence tests during childhood. (303)

13. From an early age, white parents tend to ask ______________________ questions, while black parents ask more ____________ questions. (303)

14. True or False: Fixed instructions and lack of feedback can undermine the performance of minority children on intelligence tests. (303)

15. True or False: Even spatial reasoning test performance is influenced by learning opportunities. (303)

16. What precaution should be taken when evaluating children for educational placement? (304)

____

____

Language Development

1. School-age children enlarge their vocabularies rapidly by analyzing the ________________ of complex words. (304)

2. While the word definitions of 5- and 6-year-olds are concrete, _____________ and explanations of ____________________ relationships appear in children's definitions by the end of elementary school. (304)

3. School-age children show gains in use of complex grammatical constructions such as the ______________ voice and understanding of ____________ phrases. (305)

4. School-age children have an improved ability to _________________________ _____________ in complex communicative situations. (305)

5. Cite two examples of gains in conversational strategies during middle childhood. (305)
A. ____
B. ____

6. True or False: Children who learn two languages at once in early childhood show a variety of problems with language development. (305)

7. For full development of a second language to occur, mastery must begin during ______________________. (305)

8. List the cognitive benefits of bilingualism. (305)

____

____

____

9. Bilingual education prevents ______________________, or inadequate proficiency in both languages. Cite three additional benefits of bilingual education. (306)

A. ____

B. ____

C. ____

Learning in School

1. In a traditional classroom, children are relatively (active/passive) in the learning process and are evaluated by how well they keep pace with ________________ ____________________. In an open classroom, children are relatively (active/passive) in the learning process and are evaluated in relation to ____________________. (307)

2. Children in ___________________ classrooms have a slight edge in terms of academic achievement, while those in ______________________________ classrooms are more independent, value individual differences more, and like school better. (307)

3. Despite its negative impact on motivation and learning, teachers tend to prefer the ____________________ approach for low-SES children. (307)

4. Cite three educational themes inspired by Vygotsky's sociocultural theory. (307)

A. ____

B. ____

C. ____

5. ____ is the extensive education reform based on Vygotsky's theory. Describe the main features of this model. (307-308)

____

____

____

____

6. True or False: Minority children attending KEEP classrooms perform better in school than do non-KEEP controls. (308)

7. Teachers who encourage ___________________ thinking have students who exhibit more attentiveness and higher achievement. (308)

8. Summarize ways in which teachers tend to interact differently with high-achieving as opposed to low-achieving, disruptive pupils. (308)

High-achieving: ____

Low-achieving: ____

9. How can placement in a low-ability track lead to self-fulfilling prophecies? (309)

____

____

____

10. Multigrade classrooms tend to increase _________________, _________________, and __________________. (309)

11. Explain the difference between mainstreaming and full inclusion. (309)
____
____

12. About _________ percent of school-age children have learning disabilities. ____________________________ is believed to underlie their difficulties. (309)

13. Children with special needs often do best when mainstreaming is combined with instruction by a special education teacher in a ______________________ for part of the day and when special steps are taken to promote ______________ by peers in the regular classroom. (310)

14. List three designations of giftedness among children. (310)
A. ________________________________ B. ________________________________
C. ________________________________

15. Distinguish between divergent and convergent thinking. (309)
Divergent:____
____
Convergent:____
____

16. Describe the characteristics of creative children and their families. (310)
____
____
____

17. True or False: Gifted children of all ages fare well academically and socially when offered enrichment in regular classrooms, pulled out for special instruction, or advanced to a higher grade. (311)

18. Enrichment programs that are based on Gardner's theory of multiple intelligences and include all students may help identify gifted ______________ children who are otherwise overlooked. (311)

19. True or False: One reason that American children fall behind their foreign counterparts in achievement is that their intelligence test scores are lower. (312)

20. List six cultural conditions in Japan and Taiwan that support high achievement. (312-313)
A. ____
B. ____
C. ____
D. ____
E. ____
F. ____

ASK YOURSELF . . .

Joey complained to his mother one evening that it wasn't fair that his younger sister Lizzie was almost as tall as he was. He worried that he wasn't growing fast enough. How should Rena respond to Joey's concern? (see text page 280)

___

___

___

Joshua is a South African child whose growth is stunted due to malnutrition. Maggy is an overweight Pima Indian who lives in Arizona. Explain why both Joshua and Maggy are at risk for obesity and lifelong health problems. (see text pages 282-284)

___

___

___

On Saturdays, 8-year-old Gina gathers with friends at a city park to play kickball. Besides improved ball skills, what is she learning? (see text page 289)

___

___

___

Nine-year-old Adrienne spends many hours helping her father build furniture in his woodworking shop. Explain how this experience may have contributed to Adrienne's advanced performance on Piagetian seriation problems. (see text pages 291-292)

___

___

___

One day, the children in Lizzie and Joey's school saw a slide show about endangered species. They were asked to remember as many animal names as they could. Fifth and sixth graders recalled considerably more than second and third graders. What factors might account for this difference? (see text pages 295-296)

___

___

___

___

Lizzie knows that if you have difficulty learning part of a task, you should devote most of your attention to that aspect. But she plays each of her piano pieces from beginning to end instead of picking out the hard parts for extra practice. What explains Lizzie's failure to apply what she knows? (see text page 297)

___

Desiree, a low-SES African-American child, was quiet and withdrawn while taking an intelligence test. Later she remarked to her mother, "I can't understand why that lady asked me all those questions. She's a grown-up. She *must* know what a ball and stove are for!" Using Sternberg's triarchic theory, explain Desiree's reaction to the testing situation. What interventions in the classroom and in the testing situation can enhance Desiree's academic development? (see text pages 303-304)

____

____

____

____

____

Explain how dynamic testing is consistent with Vygotsky's concept of the zone of proximal development and scaffolding. (See Chapter 5, page 160, Chapter 7, page 228, and Chapter 9, page 304.)

____

____

____

Ten-year-old Shana arrived home from school after a long day, sank into the living-room sofa, and commented, "I'm totally wiped out!" Megan, her 5-year-old sister, looked puzzled and asked, " What did'ya wipe out, Shana?" Explain Shana and Megan's different understandings of the meaning of this expression. (see text page 305)

____

____

____

Cite ways that truly bilingual education can contribute to minority children's cognitive and academic development and later-life success. (see text page 306)

____

____

____

Ricardo's first-grade teacher uses teacher-directed, whole-class instruction and has placed him in low-ability reading group. Vincente attends a mixed-grade classroom with first through third graders, and his teacher has small, heterogeneous groups of pupils work on projects in learning centers. Which child, Ricardo or Vincente, is likely to show higher achievement and to like school better? Explain. (see text pages 307-309)

____

____

____

____

Carrie is a mainstreamed first grader with a learning disability. What steps can her school and teacher take to ensure that she develops at her best, academically and socially? (see text pages 309-310)

____

____

____

____

____

SUGGESTED READINGS

Beilin, H. (1994). Jean Piaget's enduring contribution to developmental psychology. In R. D. Park, P. A. Ornstein, J. R. Rieser, & C. Zahn-Waxler (Eds.), *A century of developmental psychology* (pp. 257–290). Washington, DC: American Psychological Association. Discusses Piaget's work and influence on both past and current formulations of issues of cognitive development.

Devlin, B., Feinber, S. E., Resnick, D. P., & Roeder, K. (Eds.). (1997). *Intelligence, genes, and success: Scientists respond to The Bell Curve*. New York: Springer-Verlag. Begins by summarizing Herrnstein and Murray's arguments from their best-selling book *The Bell Curve*. Separate chapters by various experts present a scientific response to *The Bell Curve*, including reanalysis of data relied on by Hernnstein and Murray and discussion of the public policy and research implications that arise out of a more logical interpretation of the data.

Neisser, U. (Ed.). (1998). *The rising curve: Long-term gains in IQ and related measures*. Washington, DC: American Psychological Association. Leading experts in psychology, sociology, psychometrics, and nutrition present and defend different interpretations of the worldwide rise in IQ scores and the diminishing black–white gap in the school achievement of American children.

Sternberg, R. J., & Grigorenko, E. (Eds.). (1997). *Intelligence, heredity, and environment*. New York: Cambridge University Press. Provides a comprehensive summary of current theory and research on the origins and transmission of human intelligence. Discusses the investigation for genes associated with specific cognitive skills, interactionist theories, cultural relativism, educational strategies, and fallacies of earlier intelligence research.

Van der Veer, R., & Valsiner, J. (1993). *Understanding Vygotsky: A quest for synthesis*. Discusses Vygotsky's ideas in the context of Russian psychology and charts the course of Vygotsky's intellectual development. Malden, MA: Blackwell.

PUZZLE 9.1 TERM REVIEW

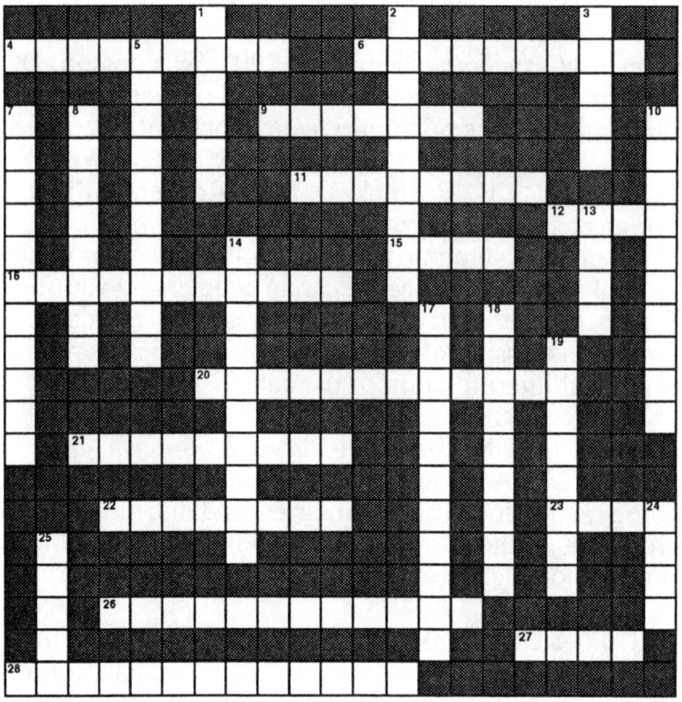

Across

4 _____ self-regulation: the process of continuously monitoring progress toward a goal, checking outcomes, and redirecting unsuccessful efforts
6 Memory strategy of repeating information
9 A greater-than-20-percent increase over average body weight, based on the child's age, sex, and physical build
11 Theory of _____ intelligences: Gardner identifies eight intelligences that permit engagement in culturally valued activities.
12 _____-streaming: placement of students with learning disabilities in regular classrooms for part of the school day
15 Classroom in which teacher shares decision making, respects different learning rates, and evaluates according to child's own performance
16 _____ classroom: children are relatively passive in the classroom; the teacher is the sole authority
20 Ability to focus on several aspects of a problem at once and relate them
21 Generation of multiple and unusual possibilities, called _____ thinking
22 _____ disability: specific disorders leading to poor achievement despite average or higher IQ
23 Education _____-fulfilling prophecy: Children may adopt teachers' attitudes toward them and start to live up (or down) to them.
26 Memory strategy of grouping related items together
27 _____ mental retardation: IQ between 55 and 70 and problems with adaptive behavior
28 Ability to mentally go through a series of steps and back again

Down

1 Children with exceptional intellectual ability
2 Ability to order items along a quantitative dimension
3 _____ skills approach to reading: emphasizes training in phonics and simplified reading materials
5 Cognitive _____: the ability to resist interference from irrelevant information
7 Concrete _____: Piaget's third stage, spanning ages 7 to 11
8 Horizontal _____: development within a stage
10 _____ interference: ability to perform seriation mentally
13 Childhood disorder involving inattention, impulsivity, and excessive motor activity (abbr.)
14 Generation of a single correct answer, called _____ thinking
17 Memory strategy of creating a relation between unrelated items
18 _____ theory of intelligence: Sternberg views processing skills, prior experience, and context as interacting to determine intelligence.
19 Nocturnal _____: nighttime bedwetting
24 _____ inclusion: placement of students with learning disabilities in the regular classroom full time
25 _____ language approach: parallels the child's natural language learning; keeps reading materials meaningful

SELF-TEST

1. Body growth in middle childhood tends to be (280)
   a. an extension of the slow, regular pattern characterized by early childhood.
   b. faster than during toddlerhood.
   c. very irregular.
   d. fast for boys, slow for girls.

2. Prolonged malnutrition (282)
   a. is often a result of children's busy schedules.
   b may lead to retarded physical growth, low intelligence test scores, poor motor coordination, inattention, and distractibility in middle childhood.
   c. does not cause permanent damage.
   d. can result from children not trying new foods.

3. Children at risk for becoming overweight are most likely from (282)
   a. low-SES families in industrialized nations.
   b. high-SES families that can afford more food.
   c. families that bottle-fed babies with cow's milk formula.
   d. families that are not athletic.

4. Over middle childhood, the frequency of injuries increases steadily, with _________ accounting for most of the rise. (285-286)
   a. male hormones and aggression
   b. sports-related injuries
   c. auto and bicycle collisions
   d. attempted suicides

5. By what age can most children print the alphabet with reasonable clarity? (286)
   a. 4
   b. 6
   c. 7
   d. 8

6. Which of the following is true about children's drawings in middle childhood? (288)
   a. It is difficult for the them to copy two-dimensional shapes.
   b. Depth cues begin to appear.
   c. Little detail is used.
   d. Drawings are usually very unorganized.

7. During middle childhood (288)
   a. boys catch up to girls in all fine motor areas.
   b. girls are behind boys in all gross motor skills.
   c. girls outperform boys in skipping, jumping, and hopping.
   d. girls can throw and kick as well as boys can.

8. Spontaneous rule-oriented games in middle childhood (289)
   a. permit children to compete with little personal risk.
   b. help children gain more mature concepts of fairness and justice.
   c. are increasing and adult-structured athletics are decreasing.
   d. do both a and b.

9. _____ is an important achievement of the concrete operational stage because it clearly shows that mental activity obeys logical rules. (291)
   a. Transitive inference
   b. Classification
   c. Conservation
   d. Horizontal décalage

10. Nadia can sort sticks of varying length into a sequence from shortest to longest. However, she cannot mentally infer that stick A is longer than stick C given that A is longer than B and B is longer than C. Which of the following abilities does Nadia lack? (291)
   a. transitive inference
   b. seriation
   c. class inclusion
   d. transformations versus states

11. Horizontal décalage refers to (292)
   a. development through the four major Piagetian stages.
   b. the tendency to focus on width rather than height in conservation tasks.
   c. gradual development within a stage.
   d. children's development of a skill across all content areas simultaneously.

12. Cross-cultural research suggests that (292)
   a. compared to non-Western societies, conservation in Western societies is greatly delayed.
   b. among the Hausa of Nigeria, the most basic conservation tasks, such as number and length, are understood as early as age 4.
   c. for children to master Piagetian concepts, they must take part in every day activities that promote this way of thinking.
   d. Hausa and American children attain conservation at about the same age.

13. Robbie Case proposes that the development of operational thinking results from (293)
   a. children's repeated use of cognitive schemes, which makes the process automatic and, in turn, frees up working memory.
   b. increases in long-term memory, which allow children to focus on combining old schemes.
   c. a sudden shift to a new stage of cognitive processing.
   d. children's interaction with adults, which allows concepts such as conservation to be socially generated.

14. Which of the following does NOT contribute to effective information processing? (293)
   a. strategy use
   b. gains in cognitive inhibition
   c. an increase in information-processing capacity
   d. an increase in memory capacity

15. As attention becomes more adaptable during middle childhood, children (294)
   a. focus on irrelevant stimuli.
   b. make planful decisions about what to attend to.
   c. attend only to familiar material.
   d. have more difficulty attending.

16. In which of the following orders do memory strategies develop? (295)
    a. organization, production, rehearsal
    b. organization, rehearsal, elaboration
    c. elaboration, organization, rehearsal
    d. rehearsal, organization, elaboration

17. Self-regulation can be enhanced by (297)
    a. simply providing feedback as to whether responses are correct or incorrect.
    b. emphasizing the value of self-correction.
    c. allowing children to discover appropriate skills without questioning their actions.
    d. allowing children to make mistakes.

18. Mathematics instruction is LEAST effective when (298)
    a. computational skills and conceptual knowledge are combined.
    b. teachers spend time encouraging girls to do well.
    c. it does not link problem-solving procedures with informally acquired understanding.
    d. it focuses on conceptual knowledge instead of drill and repetition.

19. A major shortcoming of the componential approach to the study of intelligence is that it (300)
    a. is not based on theory.
    b. regards intelligence as entirely due to causes within the child.
    c. has not generated research.
    d. does not try to uncover the underlying basis of IQ.

20. Research on African-American children adopted into middle-SES white homes during the first year of life suggests that (303)
    a. enriched environments lead to little gain in IQ, but substantial gains in academic achievement.
    b. genetic factors are largely responsible for the lower IQs of African-American children.
    c. environmental factors are largely responsible for the lower IQs of African-American children.
    d. enriched family environments overcome the effects of all other cultural influences, resulting in lasting improvement in IQ scores.

21. When asked to define "knife," Mycroft said, "A knife is something you cut with. A saw is something like a knife." Mycroft is probably (304-305)
    a. a preschooler.
    b. a first grader.
    c. a sixth grader.
    d. in high school.

22. School-age children show noticeable gains in pragmatics by (305)
    a. learning two languages at a time.
    b. using metaphors.
    c. adapting to the needs of listeners in challenging communicative situations.
    d. using 30,000 vocabulary words.

23. Which of the following statements is FALSE? (305)
    a. Bilingualism has a positive impact on cognitive development.
    b. Bilingual children are better at reflecting on language.
    c. Learning two languages in childhood tends to undermine language development.
    d. Bilingual children are advanced in analytical reasoning.

24. The dominant approach to schooling today is characterized by (307)
    a. traditional classrooms.
    b. open classrooms.
    c. KEEP classrooms.
    d. a combination of traditional and open classrooms.

25. Which of the following factors is NOT important in the superior academic performance of Asian as compared to American children? (312-313)
    a. Academic achievement is highly valued in Asian culture.
    b. Asian students are genetically superior in mathematical reasoning.
    c. Asian parents and teachers emphasize effort rather than innate ability.
    d. There is no ability grouping in Asian elementary schools; all students receive the same high-quality education.

CHAPTER 10

EMOTIONAL AND SOCIAL DEVELOPMENT IN MIDDLE CHILDHOOD

BRIEF CHAPTER SUMMARY

Erikson's stage of industry versus inferiority captures the school-age child's capacity to become productive and experience feelings of competence and mastery. During middle childhood, psychological traits and social comparisons appear in children's self-concepts, and a hierarchically organized self-esteem emerges. Gains take place in experiencing of self-conscious emotions, awareness of emotional states, and emotional self-regulation. Middle childhood also brings major advances in perspective taking. Moral understanding expands, influenced by the growing sense of personal domain.

By the end of middle childhood, children form peer groups. Friendships change, emphasizing mutual trust and sensitivity. Peer acceptance becomes a powerful predictor of current and future psychological adjustment, and the antisocial behavior of rejected children leads to severe dislike by agemates. During the school years, boys' masculine gender identities strengthen, while girls' identities become more flexible. Cultural values and parental attitudes influence these trends.

Child rearing shifts toward coregulation in middle childhood as parents grant children more decision-making power. Sibling rivalry tends to increase, and siblings often strive to be different from one another. When children experience the divorce or entry into blended families through remarriage of parents, child, parent, and family characteristics influence how well they fare. Maternal employment can lead to many benefits for school-age children, although outcomes vary depending on several child and family factors.

Fears and anxieties change during middle childhood as children experience new demands in school and begin to understand the realities of the wider world. Child sexual abuse has devastating consequences for children and is especially difficult to treat. Personal characteristics of children, a warm, well-organized home life, and social supports outside the family are related to children's ability to cope with stressful life conditions.

LEARNING OBJECTIVES

After reading this chapter, you should be able to:

10.1. Explain Erikson's stage of industry versus inferiority, noting major personality changes. (320-321)

10.2. Describe school-age children's self-concept, self-esteem, and achievement-related attributions, along with factors that affect self-evaluations in middle childhood. (321-324)

10.3. Describe changes in self-conscious emotions, emotional understanding, and emotional self-regulation in middle childhood. (324-325)

10.4. Summarize the development of perspective taking, and discuss its relationship to social behavior. (325-326)

10.5. Describe changes in moral understanding during middle childhood. (326-327)

10.6. Describe school-age children's peer groups and friendships and the contributions of each to social development. (327-329)

10.7. Describe major categories of peer acceptance, the relationship of each to social behavior, and ways to help rejected children. (329-331)

10.8. Describe changes in gender-stereotyped beliefs and gender identity during middle childhood, noting sex differences and cultural influences. (331-332)

10.9. Describe new child-rearing issues and changes in parent-child communication during middle childhood. (333)

10.10. Discuss changes in sibling relationships during middle childhood, the impact of birth order on children's experiences, and the development of only children. (333-334)

10.11. Discuss children's adjustment to divorce and blended families, noting the influence of parent and child characteristics and social supports within the family and surrounding community. (334-338)

10.12. Discuss the impact of maternal employment on school-age children's development, noting the influence of parent and child characteristics and social supports within the family and surrounding community. (338-339)

10.13. Discuss common fears and anxieties in middle childhood. (339-341)

10.14. Discuss factors related to child sexual abuse, its consequences for children's development, and ways to prevent and treat it. (341-343)

10.15. Cite factors that foster resilience in middle childhood. (43)

STUDY QUESTIONS

Erikson's Theory: Industry versus Inferiority

1. What two factors set the stage for Erikson's psychological conflict of middle childhood, industry vs. inferiority? (320)
A. ______________________________ B. ______________________________

2. List two ways in which a sense of inferiority can develop during middle childhood. (320)
A. ______________________________
B. ______________________________

Self-Development

1. List three changes in self-concept during middle childhood. (321)
A. ______________________________
B. ______________________________
C. ______________________________

2. ________________ development plays a major role in the changing *structure* of the self. The changing *content* of the self-concept is due to both ________________ capacities and ________________ from others. (321)

3. By age 6 to 7, children have at least three separate self-esteems. List them and identify which is most strongly correlated with overall self-worth. (322)
A. ________________ B. ________________ C. ________________
Strongly correlated with self-worth: ________________

4. As children adjust their self-judgments to reflect the ________________ of others, their self-esteem (rises/drops) during the early elementary school years. (322)

5. Taiwanese and Japanese children score (higher/lower) than American children in self-esteem, and they have (higher/lower) academic achievement. (322)

6. Describe child-rearing practices associated with high self-esteem in middle childhood. (322)

____

7. Mastery-oriented children attribute successes to ________________ and failures to ________________. Learned-helpless children attribute successes to ________________ and failures to ________________. (323)

8. Because they fail to make the connection between ________________ and ________________, learned-helpless children do not develop the metacognitive and self-regulatory skills necessary for success. (323)

9. What environmental variable helps account for the different attributions of mastery-oriented and learned-helpless children? (323)

____

10. True or False: Girls and low-SES ethnic minority children are especially vulnerable to learned helplessness. (323)

11. Describe attribution retraining and ways to prevent low self-esteem. (323-324)

____
____
____
____

Emotional Development

1. True or False: In middle childhood, an adult no longer has to be present for a new accomplishment to spark pride or a transgression to arouse guilt. (324)

2. List three changes in children's understanding of emotion during middle childhood. (325)
A. ________________
B. ________________
C. ________________

3. Explain how self-understanding contributes to prosocial behavior in middle childhood. (325)

____

____

4. School-age children more often mention ____________________ strategies for handling emotions than do younger children. (325)

5. Poorly regulated children are more often overwhelmed by ____________________ emotions, which interferes with _____________ behavior and peer _________________. (325)

Understanding Others: Perspective Taking

1. Match each of Selman's stages of perspective taking with the appropriate descriptions. (325-326)

_____ Understanding that third-party perspective taking can be influenced by larger societal values

_____ Recognize that self and others can have different perspectives, but confuse the two

_____ Understand that different perspectives may be due to access to different information

_____ Can imagine how the self and others are viewed from the perspective of an impartial third person

_____ Can view own thoughts, feelings, and behavior from other's perspective

1. Undifferentiated
2. Social-informational
3. Self-reflective
4. Third-party
5. Societal

2. Interventions focused on improving perspective taking are helpful in reducing ____________________ behavior and increasing ____________________ and ____________________ responding. (326)

Moral Development

1. True or False: School age children continue to depend heavily on modeling and reinforcement for engaging in good conduct. (326)

2. Trace the development of children's concepts of distributive justice during middle childhood. (326-327)
A. ____
B. ____
C. ____

3. ___________ interaction is especially important in stimulating more mature distributive justice reasoning. (327)

4. During middle childhood, children realize that moral rules and social conventions frequently _________________. (327)

5. True or False: Children in Western cultures do not use the same criteria as children in non-Western cultures to distinguish moral and social-conventional concerns. (327)

6. Explain how children's sense of the personal domain strengthens moral understanding. (327)

____

____

Peer Relations

1. List three typical characteristics of a peer group formed in middle childhood. (327)
A. ____
B. ____
C. ____

2. Describe the social skills fostered by peer group membership. (328)

____

3. How do school-age boys and girls express hostility toward the "outgroup" differently? (328)

____

____

____

4. Friendships become more ____________________ and ____________________ based, with ____________________ becoming a defining feature. (328)

5. True or False: New ideas about the meaning of friendship lead school-age children to be more selective in their choice of friends than they were at younger ages. (328)

6. Name and describe four categories of peer acceptance. (329)
A. ____
B. ____
C. ____
D. ____

7. Peer acceptance is a powerful predictor of ____________________ adjustment. (329)

8. Briefly describe the social behavior of popular children. (329)

____

____

9. Describe the social behavior of rejected-aggressive and rejected-withdrawn children. (329, 331)
Aggressive: ____

____

Withdrawn: ____

____

10. Briefly describe *peer vicitimization.* Explain how victims reinforce bullies. (330)
A.____

____

B.____

____

11. List three interventions that can help victimized children. (330)
A. ______________________________
B. ______________________________
C. ______________________________

12. Describe the social behavior and adjustment of controversial and neglected children. (331)
Controversial: ______________________________
____
Neglected: ______________________________
____

13. Describe the components of most interventions designed to help rejected children. (331)
____
____
____
____

Gender Typing

1. During the school years, children label some __________ as more typical of one gender than the other and figure out which ______________ subjects and ______________ areas are seen as "masculine" and which are labeled "feminine." (331)

2. True or False: School-age children are not open-minded about what males and females can do. (331)

3. From third to sixth grade, boys' identification with the masculine role (strengthens/declines) and girls' identification with the feminine role (strengthens/declines). (331-332)

4. Using cross-cultural evidence, show how assignment of "cross-gender" tasks can influence gender-typed behavior. (332)
____
____
____

5. True or False: Assigning boys "cross-gender" tasks, regardless of their parents' gender-typed beliefs, enhances development. (332)

Family Influences

1. ______________ works more effectively with school-age children because of their greater capacity for logical thinking. (333)

2. ________________ discipline declines over the school years. (333)

3. List the critical ingredients of coregulation. (333)
A. ______________________________
B. ______________________________
C. ______________________________
D. ______________________________

4. Siblings often try to reduce rivalry by striving to be ____________________
______________________________. (333-334)

5. The oldest sibling tends to be advantaged in ____________________
while younger siblings tend to be more ______________________________. (334)

6. Compare only versus non-only children. (334)
____
____
____

7. About ________ of American children live in single-parent households at any given time, most residing with their (mothers/fathers). (334)

8. List three factors which influence how children adjust to divorce. (335)
A. ______________________ B. ______________________
C. ______________________

9. During the period after divorce, ______________________ rises as parents try to settle disputes and mother-headed households often experience a sharp drop in ______________. Circumstances such as these often lead to " ________________ parenting." (335)

10. Briefly describe differences in the reactions of younger and older children to divorce. (335, 337)
Younger:______________________________
____
Older: ______________________________
____
____

11. Girls more often respond to divorce with ____________________ reactions, while boys in mother-custody homes are more likely to exhibit ____________________ behavior and be involved in coercive mother-child interaction. (336)

12. Most children show improved adjustment by ____ years after divorce. (336)

13. What is the overriding factor in a positive adjustment after divorce? (336)
____

14. Explain why a good father–child relationship is important for both girls and boys following divorce. (336)
Girls:______________________________
Boys:______________________________
____

15. True or False: Making the transition to a low-conflict, single-parent household is better for children than remaining in a stressed intact family. (336)

16. Describe the process of divorce mediation. (336)

____

____

17. What is necessary for the success of joint custody arrangements? (336)

____

18. List two reasons that remarriage presents difficult adjustments for most children. (337)

A. ____

B. ____

19. (Girls/Boys) adapt less favorably to a mother-stepfather custodial family. (337)

20. (Older/Younger) children find it more difficult to adjust to blended families. (337)

21. Children have more difficulty adjusting to (father-stepmother/mother-stepfather) families. (337)

22. Children of employed mothers who ______________ their work and ______________________________ to parenting show especially positive development. (338)

23. (Girls/Boys) are especially likely to benefit from maternal employment. (338)

24. Maternal employment (increases/reduces/does not change) the time children and adolescents spend with their mothers and (increases/reduces/does not change) the time they spend with their fathers. (338)

25. List three factors that help employed mothers engage in effective parenting. (338)

A. ______________________________ B. ______________________________

C. ______________________________

26. What factor is crucial in determining the effects of self-care on children. Explain. (338)

____

____

27. Low-SES children who participate in "after-care" enrichment activities show gains in ______________________________ and ______________________________. (339)

Some Common Problems of Development

1. Summarize new fears and anxieties that emerge in middle childhood. (339)

____

____

2. To what can school phobia often be traced? (339)

____

3. Describe treatments for school phobia resulting from separation anxiety and unpleasant experiences at school. (339)
Separation anxiety: ____
Unpleasant experiences: ____
____

4. List four mediating factors which influence the extent to which children are negatively affected by war. (339-340)
A. ____
B. ____
C. ____
D. ____

5. Although child sexual abuse is committed against children of both sexes, victims are more often (girls/boys). (341)

6. Describe typical characteristics of adults who engage in child sexual abuse. (341)
____
____
____
____

7. Describe adjustment problems and typical behaviors of child sexual abuse victims. (341)
____
____
____
____
____

8. Since sexual abuse usually occurs in the midst of other serious family problems, ______________________ with children and families is usually necessary. (341-342)

9. List four ways of preventing child sexual abuse. (342-343)
A. ____
B. ____
C. ____
D. ____

10. Until recently, children under age ____ were rarely asked to provide court testimony, while those age ____________ often were. Children between ____ and ____ are generally assumed competent to testify. However, children as young as age ____ have frequently served as witnesses. Explain why school-age children are better prepared to testify. (342)
____
____
____

11. Identify four approaches that are used to help children provide accurate information in court cases and reduce the trauma of testifying. (342)
A. ______________________________
B. ______________________________
C. ______________________________
D. ______________________________

12. List three broad factors that help children cope with stress and protect against maladjustment. (343)
A. ______________________________
____
B. ______________________________
____
C. ______________________________
____

ASK YOURSELF . . .

Return to page 288 of Chapter 9 and review the messages that parents send to girls about their athletic talent. On the basis of what you know about children's attributions for success and failure, why do school-age girls perform more poorly at sports than do boys, devoting less time to athletics from childhood into adulthood? (see text pages 322-324)

____
____
____
____

In view of Joey's attributions for his spelling bee performance, is he likely to enter the next spelling bee and try hard to do well? Why or why not? (see text pages 320, 322-323)

____
____
____

Joey's fourth-grade class participated in a bowl-a-thon to raise money for a charity serving children with cancer. Explain how activities like this one foster emotional development, perspective taking, and moral understanding? (see text pages 324-327)

____
____
____

Why are the strategies for regulating emotion, described on page 325, important for long-term psychological adjustment?

____
____

Apply your understanding of attributions to rejected children's social self-esteem. How are rejected children likely to explain their failure to gain peer acceptance? What impact on future efforts to get along with others are these attributions likely to have? (see text pages 322-323, 329-333)

____
____
____

Return to Chapter 8, page 266, and review the concept of androgyny. Which of the two sexes is more androgynous in middle childhood, and why? (see text pages 331-332)

"How come you don't study hard and get good grades like your sister?" a mother exclaimed in exasperation after seeing her son's poor report card. What impact do remarks like this have on sibling interaction, and why? (see text pages 333-334)

What advice would you give a divorcing couple about how to promote long-term, favorable adjustment in their two school-age sons? (see text page 336)

Nine-year-old Bobby's mother has just found employment, so Bobby takes care of himself after school. What factors are likely to affect how well Bobby fares under this arrangement? Should Bobby's mother enroll him in an "after-care" program? Explain. (see text pages 338-339)

Consider the following stressful experiences encountered by some school-age children: evaluations by parents and teachers that attribute failures to low ability; wartime trauma; and child sexual abuse. For each, cite factors that can protect children from lasting adjustment problems. What do the protective factors have in common? (see text pages 322-323, 339-340, and 341-343)

Evaluations by parents and teachers:

Wartime trauma:

Child sexual abuse:

Commonality among factors:

SUGGESTED READINGS

Bigelow, B. J., Tesson, G., & Lewko, J. H. (1996). *Learning the rules: The anatomy of children's relationships*. New York: Guilford. Presents children's understanding of the rules and rationales of social relationships with parents, siblings, peers, and teachers. Discusses a variety of relationship issues including parental authority on child compliance, sibling rivalry, close friendships, and disclosure.

Cox, M. J., & Brooks-Gunn, J. (Eds.). (1999). *Conflict and cohesion in families: Causes and consequences*. Mahwah, NJ: Erlbaum. Leading scholars working on issues of risks and resilience in families present theory and research.

Hetherington, E. M. (Ed.). (1999). *Coping with divorce, single parenting, and remarriage: A risk and resiliency perspective*. Mahwah, NJ: Erlbaum. Examines family functioning and child adjustment in different kinds of families. Discusses interactions among individual, familial, and extrafamilial risk and protective factors associated with different kinds of experiences associated with marriage, divorce, single parenting, and remarriage.

Richardson, J. T. E., Caplan, P. J., Crawford, M., & Hyde, J. S. (Eds.). (1997). *Gender differences in human cognition*. New York: Oxford University Press. Summarizes and evaluates research on gender differences in cognition. Discusses the social and cultural implications of research on sex-related differences in cognition and whether biological mechanisms account for male–female differences.

PUZZLE 10.1 TERM REVIEW

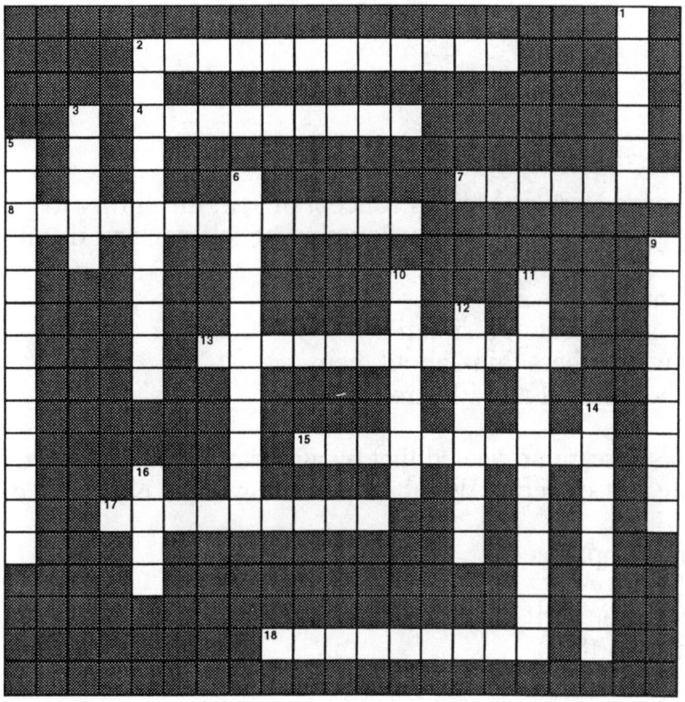

Across

2 Supervision in which parents exercise general oversight but permit children to manage moment-by-moment decisions

4 Divorce _____ is aimed at reducing family conflict by settling disputes during the period surrounding divorce.

7 Children who get many positive votes from peers

8 Children who get as many positive as negative votes from peers

13 Childre who develop learned _____ attribute success to luck and failure to low ability.

15 Rejected-_____ children engage in high rates of conflict and hostility.

17 Children who get neither positive nor negative votes from peers

18 ___-___ children look after themselves while their parents are at work. (2 words)

Down

1 Severe anxiety about attending school, called school _____

2 Social _____: children begin to judge appearance, abilities, and behavior in relation to others

3 _____ custody: an arrangement in which both parents are granted say in important decisions about a child

5 Peer _____: a form of interaction in which certain children become frequent targets of verbal and physical attacks or other forms of abuse

6 Capacity to imagine what others may be thinking and feeling, called _____ taking.

9 Rejected-_____ children are passive and socially awkward.

10 _____-oriented attributions credit success to ability and failure to insufficient effort.

11 Beliefs about how to divide up material goods fairly, called _____ justice

12 Positive outcome of Erikson's fourth stage

14 Children who get many negative votes from peers

16 _____ group: social unit with shared values and behavioral standards and a structure of leaders and follower

SELF-TEST

1. By pursuing meaningful achievement in one's culture, elementary school children show evidence of __________________ in the fourth stage of Erikson's theory. (320)
   a. initiative
   b. autonomy
   c. identity
   d. industry

2. Based on George Herbert Mead's concept of self, the ability to ___________ ___________ is critical to the development of self-concept during middle childhood. (321)
   a. become introspective
   b. focus on one aspect of a situation
   c. imagine what others think about them
   d. resolve the Oedipal and Electra conflicts

3. Susan Harter's research revealed that by age 6 to 7, children have formed at least three separate self-esteems. Which of the following is NOT one of those identified? (321)
   a. academic competence
   b. physical ability
   c. communication skill
   d. social self-worth

4. Self-esteem tends to rise for children from ____________ grades. (322)
   a. second to third
   b. third to fourth
   c. fourth to sixth
   d. sixth to eighth

5. All of the following are true about children who experience learned helplessness EXCEPT (323)
   a. they give up when tasks are difficult.
   b. they explain their failures as lack of ability.
   c. they believe that ability is a fixed characteristic of the self and cannot be changed.
   d. they usually have underlying intellectual difficulties which contribute to their feelings of incompetence.

6. Attribution retraining (323-324)
   a. is seldom successful.
   b. works well and increases IQ.
   c. works best when begun in middle childhood.
   d. has not been researched sufficiently to determine effectiveness.

7. Which of the following changes in self-conscious emotions does NOT occur during middle childhood? (324)
   a. Pride and guilt become integrated with personal responsibility.
   b. Guilt results from intentional misdeeds and not from accidental behaviors.
   c. There is recognition that individuals can experience "mixed emotions."
   d. Adults must be present for a new accomplishment to spark a sense of pride.

8. Children of ages 3 to 6 engage in (325)
   a. social-informational perspective-taking.
   b. undifferentiated perspective-taking.
   c. third-party perspective-taking.
   d. self-reflective perspective-taking.

9. Around age 6 to 7, children begin to view fairness in terms of ______. They recognize that some people deserve more rewards because they have worked harder. (326)
   a. merit
   b. equality
   c. benevolence
   d. contributory justice

10. Which of the following is true of school-age children's understanding of moral rules and social conventions? (327)
    a. They form an internal set of morals which is unrelated to conventions.
    b. They realize that situations arise in which the two overlap.
    c. They hold all people equally responsible for moral transgressions.
    d. They develop very similar beliefs despite cultural differences.

11. Which of the following is NOT true? (328)
    a. Violations of trust are regarded as serious breaches of school-age friendship.
    b. School-age children are more selective about their friendships than are preschoolers.
    c. During middle childhood, friendship is primarily a result of engaging in the same activity.
    d. Friendships become more complex and psychologically based during the school years.

12. Which of the following is not a category of peer acceptance? (329)
    a. popular children
    b. rejected children
    c. controversial children
    d. selected children

13. Which of the following is most at risk for dropping out of school and delinquency in young adulthood? (329)
    a. the neglected child
    b. the rejected child
    c. the controversial child
    d. the popular child

14. Which of the following is true? (331-332)
    a. Sex differences in activities and behaviors characteristic of Western nations are universal.
    b. Girls are more likely to experiment with "masculine" activities in cultures where the gap between males and females is wide.
    c. The sharing of domestic tasks can lead to less gender-stereotyped characteristics.
    d. Children entering elementary school are generally not aware of which academic subjects and skill areas are "masculine" and which are "feminine."

15. A transitional form of supervision in which parents exercise general oversight while permitting children to be in charge of moment-by-moment decisions is called (333)
    a. self-regulation.
    b. coregulation.
    c. authoritarian regulation.
    d. parental advisement.

16. Over middle childhood, sibling rivalry (333)
    a. tends to decrease.
    b. tends to increase.
    c. levels off.
    d. is very rare.

17. An immediate consequence of divorce that involves a disorganized family situation is called (335)
    a. optimal supervision.
    b. reconstruction.
    c. divorce mediation.
    d. minimal parenting.

18. In divorced families where the mother has custody of the children (336)
    a. girls tend to become disoriented.
    b. boys tend to become hostile and disobedient.
    c. girls tend to have greater school problems than do boys.
    d. girls tend to receive less emotional support.

19. Recent research suggests that, in the long run, divorce (336)
    a. has such harmful effects that it is rarely justified.
    b. is rarely as harmful as once believed.
    c. is better than raising children in a home where fighting prevails.
    d. is better for children than living in a home where the parents' marriage is unexciting.

20. Which of the following reconstituted family arrangements is most likely to work out well? (336-337)
    a. a boy in a mother/stepfather family
    b. a girl in a mother/stepfather family
    c. a boy in a father/stepmother family
    d. a girl in a father/stepmother family

21. Which of the following is NOT associated with the development of children whose mothers are employed? (338)
    a. the mother's work satisfaction
    b. the sex of the child
    c. the age of the child
    d. the parenting

22. Approximately _____ of school-age children develop an intense, unmanageable anxiety of some kind. (339)
    a. 5 percent
    b. 8 percent
    c. 12 percent
    d. 20 percent

23. The child sexual abuser is most likely to be (341)
    a. a mother-figure.
    b. a father-figure.
    c. an uncle or brother.
    d. a stranger.

24. When properly questioned, even ______ can recall recent events accurately, including ones that were highly stressful. (342)
    a. 3-year-olds
    b. 5-year-olds
    c. 8-year-olds
    d. 10-year-olds

25. Which of the following factors does NOT foster resiliency in middle childhood? (343)
    a. personal characteristics of the child
    b. a warm, well-organized family life
    c. social supports outside the family
    d. learned helplessness

CHAPTER 11

PHYSICAL AND COGNITIVE DEVELOPMENT IN ADOLESCENCE

BRIEF CHAPTER SUMMARY

Adolescence is a time of dramatic change leading to physical and sexual maturity. Although early theories emphasized storm and stress, recent research shows that serious psychological disturbance is not common during the teenage years. On the average, girls experience puberty 2 years earlier than do boys. In addition to heredity, nutrition and health-related factors affect maturational timing, influencing a secular trend toward earlier maturation in industrialized nations. Timing of puberty affects the psychosocial adjustment of girls and boys in opposite ways. The experience of puberty is also affected by school and cultural factors.

Puberty is related to increased moodiness and a mild rise in parent-child conflict, and is accompanied by new health concerns. Eating disorders, adolescent parenthood, sexually transmitted disease and alcohol and drug abuse are some of the most serious health concerns of the teenage years. During adolescence, both sexes improve in motor performance, with boys showing much larger gains than girls. Girls continue to receive less encouragement and recognition for athletic skills.

During Piaget's formal operational stage, adolescents become capable of abstract reasoning. Information-processing theorists agree with the broad outlines of Piaget's description of adolescent cognition. However, they refer to a variety of specific mechanisms for cognitive change and identify one, metacognition, as central to the development of abstract thought. Dramatic cognitive changes are reflected in many aspects of daily behavior, including the ability to use scientific reasoning and typical reactions such as agrumentativeness, self-concern, insensitive remarks and indecisiveness. By adolescence, boys are ahead of girls in mathematical performance. The gender gap results from both heredity and social pressures.

School transitions create new adjustment problems for adolescents. Parents, peers, and characteristics of classroom environments affect school achievement. Family background and school experience combine to influence dropping out of school, a serious problem in the United States.

LEARNING OBJECTIVES

After reading this chapter, you should be able to:

11.1. Discuss changing conceptions of adolescence over the twentieth century. (350-351)

11.2. Describe pubertal changes in body size, proportions, motor performance, and sexual maturity and the hormonal secretions that underlie them. (351-354)

11.3. Cite factors that influence the timing of puberty. (354-355)

11.4. Discuss adolescents' reactions to the physical changes of puberty, noting factors that influence their feelings and behavior. (355-356)

11.5. Discuss the impact of maturational timing on adolescent adjustment, noting sex differences and immediate and long-term consequences. (356-358)

11.6. Describe the nutritional needs of adolescents. (359)

11.7. Describe the symptoms of anorexia nervosa and bulimia, and cite factors within the individual, the family, and the larger culture that contribute to these disorders. (359-360)

11.8. Discuss social and cultural influences on adolescent sexual attitudes and behavior. (360-362)

11.9. Describe factors related to the development of homosexuality, and discuss the special adjustment problems of gay and lesbian adolescents. (362-364)

11.10. Discuss factors related to teenage pregnancy, the consequences of adolescent parenthood for development, and prevention strategies. (365-366)

11.11. Discuss the high rate of sexually transmitted disease in adolescence, noting the most common illnesses. (364-365)

11.12. Distinguish between substance use and abuse, describe personal and social factors related to each, and cite prevention strategies. (366-370)

11.13. Describe the major characteristics of formal operational thought. (370-371)

11.14. Discuss recent research on formal operational thought and its implications for the accuracy of Piaget's formal operational stage. (371-372)

11.15. Explain how information-processing researchers account for the development of abstract reasoning during adolescence. (372-374)

11.16. Describe typical reactions of adolescents that result from new abstract reasoning powers. (374-376)

11.17. Describe sex differences in mental abilities at adolescence, along with factors that influence them. (376-378)

11.18. Discuss the impact of school transitions on adolescent adjustment, and cite ways to ease the strain of these changes. (378-379)

11.19. Discuss family, peer, and employment influences on academic achievement during adolescence. (379-382)

11.20. Describe personal, family, and school factors related to dropping out, and cite ways to prevent early school leaving. (382-383)

STUDY QUESTIONS

Physical Development

Conceptions of Adolescence

1. True or False: The rate of psychological disturbance during adolescence is extremely high, supporting the conclusion that it is a period of storm and stress. (350)

2. Mead's alternative view of adolescence suggested that the ______________ ______________________ is entirely responsible for the range of teenage experiences, from erratic and agitated to calm and stress-free. (351)

3. Today we know that adolescence is a product of (biological forces/social forces/both biological and social forces). (351)

4. True or False: Adolescence, an intervening phase between childhood and full assumption of adult roles, can be found in almost all societies, but the length of adolescence varies between cultures. (351)

Puberty: The Physical Transition to Adulthood

1. Two hormones, ________________________ and _________________, contribute to the tremendous gains in body size and completion of skeletal maturation during puberty. (351)

2. The androgen __________________ leads to muscle growth, body and facial hair, and other male sex characteristics. ________________ cause the breasts, uterus, and vagina to mature and the body to take on feminine proportions. (351)

3. The rapid gain in height and weight known as the ________________________ is underway for American girls just after age ____, for boys around age _____. Growth in overall body size is complete for most girls by age ______ and for boys at age _____. (352)

4. True or False: During adolescence, the cephalocaudal trend of growth reverses. (352)

5. While girls add more _______________ during adolescence, boys add comparatively more ________________. (352)

6. Adolescents get an average of ____ to _____ hours of sleep at night, a(n) (increase/decrease) from middle childhood. (353)

7. Describe the motor development of adolescent girls and boys. What affect do the differences have on physical education in junior high school? (353)
Adolescent girls: ____
Adolescent boys: ____
Affect on physical education: ____
____

8. Sports and exercise influence ______________________ and ___________________ development, as they provide important lessons in ________________________, _________________________, __________________________, and _________________________. (353-354)

9. Female puberty usually begins with __________________________________ and __________________________. Menarche typically occurs around ______ years of age among American adolescents, although it varies from 10 1/2 to 15 1/2 years. (354)

10. The first sign of puberty in boys is enlargement of the ____________. (354)

11. The growth spurt occurs (earlier/later) in the sequence of pubertal events for boys than girls. (354)

12. Spermarche typically occurs around age ______. (354)

13. List two factors that appear to be responsible for individual differences in pubertal growth. (355)
A. _______________________________ B. _________________________________

14. As nutrition and health increased from 1860 to 1970 in industrialized nations, the average age of menarche (declined/increased), demonstrating the role of physical well-being in adolescent growth. (355)

The Psychological Impact of Pubertal Events

1. List three related factors that affect girls' feelings about menarche. (355)
A. ____
B. ____
C. ____

2. True or False: Girls whose fathers are told about pubertal changes adjust especially well. (355)

3. Overall, boys get much (more/less) social support for the physical changes of puberty. (356)

4. Unlike industrialized nations, many tribal and village societies formally recognize pubertal changes by celebrating with a _________________________ marking changes in privilege and responsibility. (356)

5. The absence of an accepted marker of maturity can make the transition to adulthood confusing, since in some contexts adolescents are treated as ____________ and in others they are still treated as ______________. (356)

6. _______________ factors appear to combine with ________________ influences to increase teenagers' moodiness. (356)

7. ________________________ between parents and children, exemplified by increased conflict, may be a modern substitute for the physical departure seen in nonhuman primates and nonindustrialized nations. (357)

8. True or False: The increase in parent–child conflict at adolescence is generally very extreme. (357)

9. Early maturation has social benefits for (boys/girls) and late maturation is socially beneficial for (boys/girls). (357)

10. List two possible reasons for these maturational timing effects. (357)
A. ____
B. ____

11. Explain how school contexts can modify maturational timing effects. (358)
____
____
____

12. True or False: When long-term outcomes are examined, many of the effects of maturational timing on well-being appear to reverse themselves. (358)

Health Issues

1. Rapid body growth leads to a dramatic rise in ________________. (359)

2. True or False: Of all age groups, the eating habits of adolescents are the poorest. (359)

3. The most common nutritional problem of adolescence is ____________________.
List three other common nutritional deficiencies of the teenage years. (359)
A. ____________________ B. __________________ C. ________________

4. Anorexics have an extremely distorted _________________, seeing themselves as fat even when they are dangerously thin. (359)

5. Cite cultural, individual, and familial factors related to anorexia nervosa. (359-360)
Cultural: ____
Individual: ____
Familial: ____
____
____

6. Describe two approaches to treating anorexia nervosa. Circle the letter of the one that is the most successful. (360)
A. ____________________________ B. ____________________________

7. Bulimia is (more/less) common than anorexia nervosa. (360)

8. Bulimics share many characteristics with anorexics, but differ in their lack of ________________________ in many areas of their lives. Bulimia (is/is not) easier to treat than is anorexia. (360)

9. American society falls on the (permissive/restrictive) side of the cultural continuum of attitudes toward adolescent sexuality. (360)

10. Explain why the messages teenagers get from parents and television are confusing. (360-361)

____

____

11. Sexual attitudes and behavior of adolescents have become (more/less) liberal over the past 30 years. (361)

12. True or False: Most sexually active teenagers engage in sexual relations with only one partner at a time and engage in relatively low levels of sexual activity. (361)

13. Summarize personal, family, peer, and educational variables that are linked to sexual activity during adolescence. (361)

Personal: ____

Family: ____

Peer: ____

Educational: ____

14. Some teenagers fail to use birth control because advances in perspective taking make them extremely concerned with ______________________________, because they overlook the ____________________ of risky behaviors when under social pressure, and because of limited access to information and parental support. (362)

15. Identify and briefly describe each of the three phases that homosexual adults and adolescents move through in coming out to themselves and others. (362-363)

A. ____

____

B. ____

____

C. ____

____

16. Briefly explain how heredity might lead to homosexuality. (363-364)

____

____

____

17. List two stereotypes or misconceptions about homosexuality. (364)

A. ____

B. ____

18. True or False: Adolescents have the highest rates of sexually transmitted disease (STD) of any age group. (364)

19. True or False: Nearly all of the cases of AIDS that appear in young adulthood originate in adolescence. (365)

20. The adolescent pregnancy rate in the United States is much higher than that of European nations. List three ways in which the United States differs from these nations. (365)

A. ____

B. ____

C. ____

21. True or False: The United States has one of the highest adolescent abortion rates of any developed country. (365)

22. List two reasons why becoming a parent is more problematic for adolescents than for adults. (366)
A. ______
B. ______

23. Describe how the lives of pregnant teenagers worsen after the baby is born in the following three respects: (366)
Educational attainment: ______
______
Marital patterns: ______
______
Economic circumstances: ______
______

24. Babies of teenage mothers are more likely to experience ______ ______ and are at risk for poor ______. (366)

25. Cite examples of rearing conditions that appear to be linked to adolescent parenthood. (367)
______
______

26. The development of children born to teenage mothers, who do not repeat the pattern of early childbearing, (is/is not) compromised. (367)

27. Knowledge about ______ and ______ is crucial to preventing teenage pregnancy. However, sex education must help teenagers to bring together what they ______ with what they ______. (367)

28. More effective sex education programs emphasis the following three elements: (367-368)
A. ______
B. ______
C. ______

29. True or False: Social competence is important in preventing adolescent pregnancy and parenthood. (368)

30. True or False: In European countries where contraception is easily available to teenagers, sexual activity is not higher than in the United States, but pregnancy and abortion rates are much lower. (368)

31. Describe interventions that are necessary to help adolescent parents and their babies. (368)
______
______
______
______

32. True or False: Teenage alcohol and drug abuse in the United States is higher than in any other industrialized nation. (369)

33. True or False: Teenagers who experiment with alcohol, tobacco, and marijuana are seriously maladjusted. (369)

34. Drug abusers are seriously troubled adolescents who express unhappiness through ______________________ acts, have a history of family problems, and have friends who ____________________________________. (369)

35. Over time, adolescent substance abusers fail to learn responsible ________ __________ skills and alternative ___________ strategies and therefore fail much more often in occupational and interpersonal pursuits. (369)

36. Cite three ways schools can help reduce drug experimentation in adolescence. (369)
A.____
B.____
C.____

37. How does drug abuse prevention differ from prevention of occasional use. (370)
____
____

Cognitive Development

Piaget's Theory: The Formal Operational Stage

1. In formal operational reasoning, ______________ objects and events are no longer necessary for logical thought: adolescents can come up with general logical rules through _______________ reflection. (370)

2. Name and describe two major features of formal operational reasoning. (370-371)
A.____
____
____
B.____
____

3. Piaget acknowledged that ________________ plays a larger part in thinking during adolescence. (371)

4. True or False: School-age children have the capacity for hypothetico-deductive reasoning and propositional thought. However, it is inferior to that of adolescents in depth, complexity, and abstractness. (372)

5. People are more likely to think abstractly in situations with which they have had __________________________________. (372)

An Information-Processing View of Adolescent Cognitive Development

1. Describe how each of the following mechanisms changes in adolescence. (372)
Attention:____
Strategy Use: ____
Knowledge: ____
Metacognition:____
Processing Capacity: _____________________________________
Which mechanism do researchers regard as central to the development of abstract thought? ________________________

2. True or False: As individuals move from childhood to adolescence, they are better able to distinguish theory from evidence and use logical rules to examine their relationship. (373)

3. Identify three factors that support adolescents' skill at coordinating theory with evidence. (373)
A. ____
B. ____
C. ____

Consequences of Abstract Thought

1. Improved formal operational abilities contribute to greater __________________ in adolescence. As long as they remain focused on ____________________, parent-child disagreements promote development. (374)

2. Describe two distorted images of the relation between self and others that appear at adolescence. (374-375)
Imaginary audience: ____
____
Personal fable:____
____

3. When are these two distorted images the strongest? (375)
____

4. Because abstract thinking permits adolescents to go beyond the real to the possible, ______________________ increase. (375)

5. Parents can help teenagers balance the ideal and the real by ______________________ ______________________ while ____. (377)

6. While adolescents become much better at __________________________ when given a homework assignment, they often feel overwhelmed by planning and decision-making in daily life. (377)

Sex Differences in Mental Abilities

1. True or False: Girls and boys differ in general intelligence. (377)

2. Males' superior spatial skill performance is apparent in ________________________ and ________________________ tasks, but not in ________________________ tasks. (377)

3. Both ____________________ and __________________ contribute to the sex differences in spatial and math performance. (377)

4. List three social factors related to girls' poorer performance in math. (376-377)
A. ____
B. ____
C. ____

5. List three ways in which girls' interest in and confidence at doing math and science can be promoted. (378)
A. ____
B. ____
C. ____

Learning in School

1. With each school transition in adolescence, course grades ____________. (378)

2. The (earlier/later) the school transition occurs, the more dramatic and long-lasting its impact on psychological well-being, especially for ________. (378)

3. List three ways of helping adolescents adjust to school transitions. (379)
A. ____
B. ____
C. ____

4. ________________ parenting is linked to achievement in adolescence, as is parents' ________________ with the school. (379)

5. Explain how Asian and African-American child-rearing styles influence achievement in adolescence. (380)
Asian: ____
____
African-American: ____
____

6. Teenagers whose parents value achievement are likely to choose friends who (share/oppose) those values. (380)

7. How can adults help motivate adolescents' academic achievement in the face of peer pressure against doing well in school? (381)
____
____

8. Many adolescents report that their classes lack ______________________________, a circumstance which dampens their motivation. (381)

9. True or False: African-American and Hispanic students now, on average, achieve mastery of reading, writing, math, and science equal to that of white children. (381)

10. True or False: Mixed-ability rather than tracked classes are desirable into the early years of secondary school. (381)

11. True or False: A good student from a disadvantaged family is just as likely to end up in an academically oriented, college-bound track as a good student from a higher-SES background. (381)

12. Educational decisions are (more/less) fluid in the United States than in other industrialized nations. Still, by adolescence, ________________________ differences in the quality of education and achievement have already drastically sorted American students. (381)

13. Students who work more than _____ hours per week have poorer school attendance, lower grades, and less time for extracurricular activities. (382)

14. Work-study programs are related to (positive/negative) school and work attitudes and (improved/declining) achievement for low-SES students at risk for dropping out. (382)

15. By age 18, ____ percent of American young people drop out of high school. (382)

16. Teenagers who drop out of school feel __________________ from school life and have parents who are ______________________ in their education and respond with ________________________ when their children bring home poor grades. (383)

17. List four successful approaches to preventing early school leaving. (383)

A. ____
B. ____
C. ____
D. ____

18. True or False: Over the last half-century, the percentage of adolescents completing high school has more than doubled. (383)

ASK YOURSELF . . .

Millie, mother of an 11-year-old son, is convinced that the rising sexual passions of puberty cause rebelliousness. Where did this belief originate? Explain why it is incorrect. (see text page 350)

____
____
____
____

Explain why early maturing girls, as opposed to early maturing boys, often experience adjustment problems. Cite school and cultural factors that affect the development of early maturing girls--during adolescence and adulthood. (see text pages 357-358)

How might adolescent moodiness contribute to the psychological distancing between parents and children that accompanies puberty? (Hint: Think about bidirectional influences in parent–child relationships discussed in previous chapters.) (see text pages 356-357)

Fourteen-year-old Lindsay says she couldn't possibly get pregnant because her boyfriend would never do anything to "mess her up." What factors might account for Lindsay's unrealistic reasoning? (see text page 361)

At age 16, Veronica gave birth to her first child. When at age 17 she had a second, her parents told her they didn't have room for two babies. Veronica dropped out of school and moved in with her boyfriend, Todd. A few months later, Todd left because he couldn't stand being tied down with the babies. Why are Veronica and her children likely to experience long-term hardships? (see text page 366)

Explain how adolescent substance abuse follows the pattern of other teenage health problems in being the product of both psychological and social forces. (see text pages 368-369)

Cassie insisted that she had to have high heels to go with her prom dress. "No way I can wear these low heels, Mom. They'll make me look way too short next to Louis, and the whole evening will be ruined!" Why is Cassie so concerned about a detail of her appearance that most people would be unlikely to notice? (see text pages 374-375)

Adolescent idealism and criticism, although troublesome for parents, are beneficial in the long run, both for the developing individual and society. Explain why this is so. (see text pages 375-376)

Research shows that girls perform more poorly than boys on certain formal operational tasks, such as the pendulum problem. On the basis of what you know about the development of formal operational thought and sex differences in cognitive abilities, how would you account for this finding? (see text pages 371-372, 376-377)

What steps can be taken to increase girls' interest in taking math and sciences? What impact are these efforts likely to have on lifelong development? (see text pages 376-378)

Tanisha is finishing sixth grade. She could either continue in her current school through eighth grade or switch to a much larger junior high school. What would you suggest she do, and why might her decision have lasting consequences for her academic development and educational attainment? (see text pages 378-379)

In a workshop for parents of adolescents, one father asks what he might do to encourage his teenage children to do well in school. Provide a list of suggestions along with reasons each is effective. (see text pages 379-380)

SUGGESTED READINGS

Burt, M. R., Resnick, G., & Novick, E. R., (1998). *Building supportive communities for at-risk adolescents: It takes more than services.* Washington, DC: American Psychological Association. Profiles nine outstanding youth-serving programs in the United States, which provide a mix of educational, counseling, recreational, vocational, and leadership activities.

Johnson, N. G, Roberts, M. C., & Worell, J. P. (Eds.). (1999). *Beyond appearance: A new look at adolescent girls.* Washington, DC: American Psychological Association. An edited volume that explores key topics to understanding girls' adolescent development, including gender-role behaviors, body image issues, relationships with family and friends, sexual decision making, and school- and community-based experiences.

Schulenberg, J., Maggs, J., and Hurrelmann, K. (Eds.). (1997). *Health risks and developmental transitions during adolescence.* New York: Cambridge University Press. Examines how various developmental transitions associated with adolescence afford risks and opportunities for young persons' mental and physical health.

Siegel, A. W., Cuccaro, P., Parsons, J. T., Wall, J., & Wienberg, A. D. (1993). Adolescents' thinking about emotions and risk-taking. In J. M. Puckett & H. W. Reese (Eds.), *Mechanisms of everyday cognition.* Hillsdale, NJ: Erlbaum. Provides insight and research regarding the everyday cognitive world of the adolescent. Both adolescent and adult egocentrism are used as frameworks for demonstrating and explaining how teenagers think about their social surroundings.

Wilson, D. K., Rodrigue, J. R., & Taylor, W. C. (Eds.). (1997). *Health-promoting and health-compromising behaviors among minority adolescents.* Washington, DC: American Psychological Association. Explores specific health-compromising and health-promoting behaviors of minority youths that contribute to their increased risk for numerous health problems.

PUZZLE 11.1 TERM REVIEW

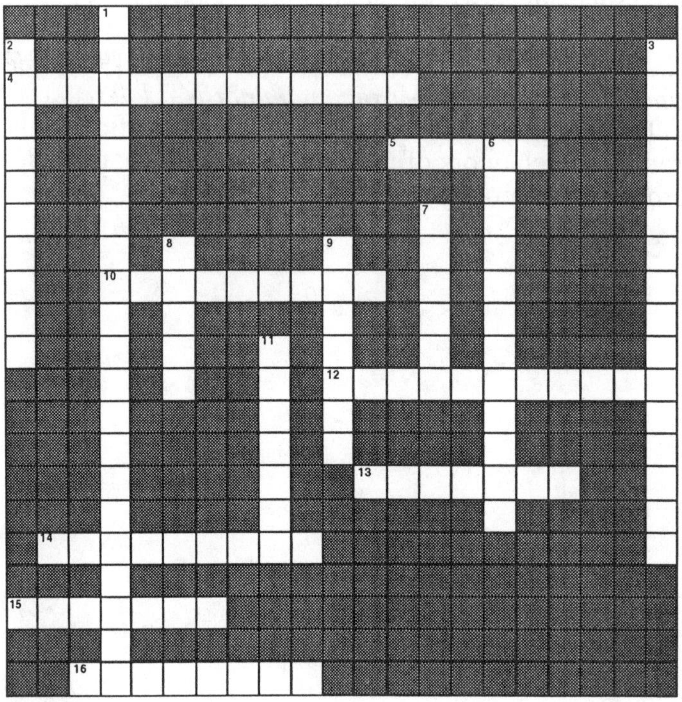

Across

4 _____ thought: evaluating the logic of verbal statements without referring to real-world circumstances

hypotheses

5 Body _____: conception of and attitude toward one's physical appearance

10 _____ audience: adolescents' belief that they are the focus of everyone's attention

12 A period in which individuals cross from childhood to adulthood

13 Biological changes leading to an adult-sized body and sexual maturity

14 _____ sex characteristics: serve as signs of sexual maturity, but do not involve the reproductive organs

15 Eating disorder in which individuals go on eating binges by purging methods

16 First menstruation

Down

1 _____ reasoning begins with a theory of all possible factors that could influence an outcome and deduction of specific

which are tested systematically. (2 words)

2 First ejaculation of seminal fluid

3 Eating disorder in which individuals starve themselves because of a compulsive fear of getting fat (2 words)

6 Rapid gain in height and weight (2 words)

7 _____ operational stage: Piaget's final stage

8 Personal _____: adolescents' belief that they are special and unique

9 _____ sexual characteristics: those which involve the reproductive organs directly

11 _____ trends in physical growth: changes in body size and rate of growth from one generation to the next

SELF-TEST

1. The first researcher to point out that social and cultural factors contribute to adolescent adjustment was (351)
   a. Vygotsky.
   b. Freud.
   c. M. Mead.
   d. G. S. Hall.

2. Adrenal androgens (351)
   a. are not present in girls.
   b. affect primary sex characteristics in boys.
   c. influence girls' growth spurt.
   d. have a greater impact on boys than on girls.

3. When does menarche occur relative to the growth spurt? (354)
   a. before the growth spurt begins
   b. during the early part of the growth spurt
   c. during the latter part of the growth spurt
   d. after the growth spurt has ended

4. Which of the following statements is true? (354)
   a. A secular trend toward later menarche exists in industrialized nations.
   b. The secular trend in pubertal timing is believed to be due to genetic factors.
   c. Secular gains have dramatically increased in industrialized nations in recent years.
   d. A leveling-off of secular trends in many industrialized nations has occurred in recent years.

5. Compared to school-age children and adults, adolescents (356)
   a. experience moodiness due primarily to hormones.
   b. have stable moods.
   c. experience lower moods.
   d. experience low points during self-chosen leisure.

6. Which of the following are likely to have positive self-images in adolescence? (357)
   a. late maturing boys and girls
   b. early maturing boys and girls
   c. late maturing boys and early maturing girls
   d. early maturing boys and late maturing girls

7. Which of the following statements is true? (359-360)
   a. Anorexia nervosa is more common than bulimia.
   b. Girls with anorexia nervosa feel extremely guilty about their eating habits, whereas girls with bulimia do not.
   c. Bulimia is prevalent in low-SES families, whereas anorexia is prevalent in middle-SES families.
   d. Bulimia is usually easier to treat than is anorexia nervosa.

8. Which of the following is NOT true? (360-361)
   a. Early sexual activity is more common among early physically maturing teenagers.
   b. The rate of teenage sexual activity in the United States is about the same as in Western European nations.
   c. The rate of premarital sex among young people is continuing to rise.
   d. Early sexual activity is more common among teenagers whose parents are divorced.

9. Which of the following statements is NOT true? (363-364)
   a. Heredity makes a contribution to homosexuality.
   b. Some heterosexual adolescents are attracted to members of the same sex.
   c. There are very few stereotypes about homosexuality.
   d. About 3 to 6 percent of teenagers discover that they are lesbian or gay.

10. Teenagers in greatest danger of being infected by an STD are (364-365)
   a. poverty-stricken teenagers who feel a sense of hopelessness.
   b. middle-income teenagers whose parents are permissive.
   c. middle-income teenagers whose parents have high expectations.
   d. all sexually active teenagers are equally susceptible.

11. Adolescent mothers (365-366)
   a. almost always go back to school.
   b. often experience prenatal and birth complications.
   c. are less likely to divorce than their peers.
   d. know an extensive amount about child development.

12. Long-term consequences of adolescent drug abuse include all of the following EXCEPT (369)
   a. failure to learn responsible decision-making skills.
   b. depression and antisocial behavior.
   c. entering into marriage and childbearing prematurely.
   d. development of alternative coping patterns.

13. When faced with a problem, adolescents begin with a general theory, deduce from it specific predictions about outcomes, and then test the predictions—a process Piaget called (371)
   a. hypothetico-deductive reasoning.
   b. horizontal décalage.
   c. propositional thought.
   d. inductive reasoning.

14. Formal operations seem to be (372)
   a. an abrupt development.
   b. due to children's independent efforts to make sense of their world.
   c. universal and emerging in all contexts at once.
   d. gradual and situation- and task-specific.

15. Which of the following mechanisms do researchers consider central to the development of abstract thought? (372)
   a. metacognition
   b. attention
   c. processing capacity
   d. strategy use

16. Argumentativeness during adolescence seems to be due to (374)
    a. general disagreeableness.
    b. the exercising of new reasoning powers.
    c. intense self-reflection.
    d. improvements in self-regulation.

17. The personal fable contributes to (375)
    a. risk-taking.
    b. disengagement from others.
    c. poor school work.
    d. peer conformity.

18. Teenage idealism often leads to a disparity between adults and teens called (375)
    a. the personal fable.
    b. egocentrism.
    c. the generation gap.
    d. social commitment.

19. Adolescents' planning and decision making is (375)
    a. more effective in daily life than in school work.
    b. more effective in school work than in daily life.
    c. very effective in both school work and daily life.
    d. poorly developed in both school work and daily life.

20. Sex differences in mental ability appear in all of the following except (377)
    a. spatial visualization tasks.
    b. math performance.
    c. mental rotation tasks.
    d. spatial perception tasks.

21. In general, research shows that the earlier the school transition (378)
    a. the more powerful and long-lasting its impact, especially for boys.
    b. the more powerful and long-lasting its impact, especially for girls.
    c. the less serious its impact, especially for boys.
    d. the less serious its impact, especially for girls.

22. Which of the following is an effective way to ease the strain of school transition in adolescence? (379)
    a. reorganizing schools into 6-3-3 arrangements
    b. making sure that academic expectations in junior high are tougher than those in elementary school
    c. creating smaller social units within larger schools
    d. reducing the number of extracurricular activities available in high school

23. A factor that supports high achievement during adolescence is (379-380)
    a. authoritative parenting.
    b. a strict, rule-oriented way of reasoning.
    c. peer encouragement to ease up on school assignments and relax.
    d. departmentalized school organizations.

24. Students in low-ability groups (381)
    a. often get poor quality instruction.
    b. show gains in achievement, since instruction is adapted to their needs.
    c. are viewed as successful by their peers.
    d. like school better than students in mixed-ability classes.

25. Potential dropouts benefit from (383)
    a. remedial instruction in small classes.
    b. less job-related instruction.
    c. larger classes.
    d. large high schools.

CHAPTER 12

EMOTIONAL AND SOCIAL DEVELOPMENT IN ADOLESCENCE

BRIEF CHAPTER SUMMARY

Erikson's stage of identity versus identity confusion recognizes the formation of a coherent set of values and life plans as the major personality achievement of adolescence. The development of an organized self-concept and more differentiated sense of self-esteem is affected by several factors, including abstract reasoning, and family, school, and community contexts. Changes in self-concept and self-esteem prepare the young person for constructing an identity. Adolescents vary in their degree of progress toward developing a mature identity. Identity achievement and moratorium are adaptive statuses associated with positive personality characteristics. Teenagers who remain in identity foreclosure or identity diffusion tend to have adjustment difficulties.

According to Kohlberg, morality changes from concrete, externally controlled reasoning in late childhood to more abstract, principled justifications for moral choices in adulthood. A broad range of family, school, peer, and cultural factors foster moral development. As individuals advance through Kohlberg's stages, moral reasoning becomes more closely related to behavior.

Biological, social, and cognitive forces combine to make early adolescence a period of gender intensification. Over the adolescent years, relationships with parents and siblings change as teenagers strive to establish a healthy balance between connection to and separation from the family. As adolescents spend more time with peers, intimacy and loyalty become central features of friendship. Adolescent peer groups are organized into tightly knit groups called cliques, and as teenagers become interested in dating, several cliques come together to form a crowd. Although peer pressure rises in adolescence, most teenagers do not blindly conform to the dictates of agemates.

Although most young people move through adolescence with little difficulty, a few encounter major disruptions in development. Depression is the most common psychological problem of the teenage years, influenced by a diverse combination of biological and environmental factors. The suicide rate increases dramatically at adolescence. Many teenagers become involved in some delinquent activity, but only a few are serious or repeat offenders. Family, school, peer, and neighborhood factors are related to delinquency.

LEARNING OBJECTIVES

After reading this chapter, you should be able to:

12.1. Discuss Erikson's account of identity development. (390-391)

12.2. Describe changes in self-concept and self-esteem during adolescence. (391-392)

12.3. Describe the four identity statuses, their relationship to psychological well-being, and factors that affect identity development. (392-394)

12.4. Describe Piaget's theory of moral development and Kohlberg's extension of it, and evaluate the accuracy of each. (394-398)

12.5. Describe environmental influences on moral reasoning. (399)

12.6. Evaluate claims that Kohlberg's theory does not adequately represent the morality of females, and describe the relationship of moral reasoning to behavior. (400)

12.7. Explain why early adolescence is a period of gender intensification. (401)

12.8. Discuss changes in parent–child and sibling relationships during adolescence. (401-403)

12.9. Discuss changes in friendships and peer groups during adolescence, and describe the contributions of each to emotional and social development. (403-405)

12.10. Describe adolescent dating relationships, and discuss conformity to peer pressure in adolescence. (405-406)

12.11. Discuss factors related to adolescent depression and suicide along with approaches to prevention and treatment. (406-410)

12.12. Discuss factors related to delinquency and ways to prevent and treat it. (410-412)

STUDY QUESTIONS

Erikson's Theory: Identity versus Identity Confusion

1. Summarize what is involved in constructing an identity. (390)

____

____

2. True or False: According to Erikson, resolution of psychological conflicts during earlier stages of development has nothing to do with constructing an identity at adolescence. (390)

3. According to Erikson, an ____________________ is a temporary period of confusion and distress as adolescents experiment with alternatives before settling on values and goals. Do current theorists still view this process as a "crisis?" __________ (390)

Self-Development

1. List four ways in which self-concept during middle and late adolescence differs from that of late childhood and early adolescence. (391)

A. ____

B. ____

C. ____

D. ____

2. Self-esteem continues to ____________________ during adolescence. Except for a temporary decline associated with ________________________, self-esteem rises for most adolescents. (391)

3. List five factors related to sense of self-esteem in adolescence. (391-392)
A. ____________________ B. ____________________ C. ____________________
D. ____________________ E. ____________________

4. Match each of the following forms of identity status with the appropriate description. (392)

____	Committed to values and goals without taking time to explore alternatives	1. Identity achievement
____	Have not yet made definite commitments and are still exploring	2. Moratorium
____	Committed to self-chosen values and goals after having already explored alternatives	3. Identity foreclosure
____	Lack clear direction, are not committed to values and goals, and are not actively seeking them	4. Identity diffusion

5. Many adolescents start out as identity ________________ and ________________, but by late adolescence, they have moved toward ________________ and ____________________________________. (392)

6. True or False: Girls show more sophisticated reasoning in identity areas related to sexuality and family versus career options. (391)

7. Summarize the personality characteristics associated with the following identity statuses. (393)
Identity achievement and moratorium:____
____
Identity foreclosure: ____
____
____
Identity diffusion: ____
____
____

8. Match the following forms of identity status with the appropriate descriptions of associated cognitive, family, school, community, and larger cultural factors. (Descriptions can apply to more than one identity status.) (393-394)

_____	Assume that absolute truth is always attainable	1. Identity achievement
_____	Lack confidence in ever knowing anything with certainty	
_____	Appreciate that rational criteria can be used to choose among alternative visions	2. Moratorium
_____	Feel attached to parents but free to voice opinions	3. Identity foreclosure
_____	Have close bonds with parents but few opportunities for separation	
_____	Report low levels of warmth and openness at home	4. Identity diffusion
_____	Fostered by high-level thinking, extracurricular and community activities, taking on of responsible roles, and vocational training in school	

9. Discrimination and inequality interfere with ethnic minority adolescents' ability to develop a healthy ________________. (395)

10. List four ways in which minority adolescents can be helped to resolve identity conflicts constructively. (395)
A. ____________________
B. ____________________
C. ____________________
D. ____________________

11. Biculturally identified adolescents tend to have a higher sense of ____________, a greater sense of ____________, and a more positive ____________. (395)

Moral Development

1. As people realize others can have different perspectives on moral matters and that intentions, not just outcomes, should serve as the basis for judging behavior, they are transitioning to the stage of ____________ morality. (396)

2. When adolescents use reciprocity, they express the same concerns for the welfare of ____________ as they do for ____________. (396)

3. True or False: Although not entirely correct, Piaget's account of morality does describe the general direction of moral development. (396)

4. In Kohlberg's moral dilemmas, it is the ________ an individual reasons, rather than the ____________ of the response, which determines moral maturity. (396)

5. On what two factors did Kohlberg believe moral reasoning was based? (396-397)
A. ____________________
B. ____________________

6. Match each of the following moral orientations with the appropriate description. (397-398)

____ Emphasis is on fair procedures for changing laws to protect individual rights and majority needs
____ Fear of authority and avoidance of punishment motivate morality
____ Abstract universal principles valid for all humanity guide moral decisions
____ Morality is used to satisfy personal needs
____ Moral conformity is justified by a duty to uphold laws and rules
____ The desire to maintain the affection and approval of others motivates morality

Preconventional Level:
1. Punishment and obedience
2. Instrumental purpose

Conventional Level:
3. "Good boy-good girl"
4. Social-order-maintaining

Postconventional Level:
5. Social contract orientation
6. Universal ethical principle

7. Moral development is very slow and gradual. Stage 1 and 2 reasoning decline in ____________, whereas stage 3 increases through ____________ and then declines. Stage 4 reasoning rises over ____________ and is the typical response in adulthood. Few people move beyond stage ______. (398)

8. Gilligan believes that feminine morality emphasizes an ________________ that is devalued in Kohlberg's model. (399)

9. True or False: Research shows that when given moral dilemmas, females fall behind males in development according to Kohlberg's scheme. (399)

10. What three behaviors on the part of parents are related to the most gains in teenagers' moral development? (399)
A. ________________________________ B. ________________________________
C. ________________________________

11. True or False: Years of schooling is one of the most powerful predictors of moral maturity. (399)

12. True or False: Spontaneous moral conflicts between peers promote moral understanding. (399-400)

13. True or False: Early participation in societal institutions promotes moral development. (400)

14. Kohlberg's theory (does/does not) capture all of the important aspects of moral thinking in every culture. (400)

15. Kohlberg believed that moral thought and action come closer together at the ________________ levels of morality. Is this belief supported by research? ________ (400)

Gender Typing

1. Gender intensification is greater in adolescent (girls/boys) than (girls/boys). (401)

2. List four reasons for adolescent gender intensification. (401)
A. ____
B. ____
C. ____
D. ____

3. True or False: Androgynous adolescents tend to be psychologically healthier. (399)

The Family

1. ________________ parenting continues to be the most effective style during adolescence. (401-402)

2. Once teenagers ____________________ their parents, they no longer bend as easily to parental authority. (402)

3. Parents who are financially secure, invested in their work, and content with their marriages usually find it (easier/harder) to grant their teenagers an appropriate degree of autonomy. (402)

4. List six family circumstances that pose challenges for adolescents. (402-403)
A.______________________________ B.______________________________
C.______________________________ D.______________________________
E.______________________________ F.______________________________

5. Teenage sibling relationships become (more/less) intense in both positive and negative feelings. In addition, they invest (more/less) time in their relationships.(403)

Peer Relations

1. Cite the two characteristics of friendship stressed by teenagers. (403-404)
A. ______________________________ B. ______________________________

2. Girls are more likely to mention _______________ and _______________ in their descriptions of friendships. Androgynous boys (are/are not) just as likely as girls to form intimate same-sex friendships. (404)

3. List three ways in which adolescent friendships promote emotional and social development. (404)
A.____
____
B.____
____
C.____
____

4. __________________, ____________________, and __________________ factors are all important in determining what cliques and crowds teenagers belong to. (404)

5. Positive peer impact is greatest for teenagers whose own parents are ____________________, whereas negative peer impact is strongest for teenagers whose parents ______________________________ child-rearing styles. (405)

6. List the vital functions that cliques and crowds serve. (405)
Cliques: ____
____
Crowds: ____
____

7. Describe younger and older adolescents' different reasons for dating. (405)
Younger: ____
____
Older: ____
____

8. List two unique challenges that homosexual youth's face in dating. (406)
A. ______________________________ B.______________________________

9. Besides fun and enjoyment, list three benefits of dating given that it does not begin too soon. (406)
A. ______________________________ B. ______________________________
C. ____

10. Describe the differing spheres of influence of parents and peers during adolescence. (406)
Parents: ______________________________
Peers: ______________________________

11. What type of child rearing is related to greater resistance to unfavorable peer pressure during adolescence? (406)

____

Problems of Development

1. About _________ percent of teenagers experience one or more depressive episodes and about _____ percent are chronically depressed. (407)

2. Explain why adolescent depressive symptoms are likely to be overlooked by parents and teachers. (407)
A. ______________________________
B. ______________________________

3. True or False: Depression in adolescence is caused by a combination of biological and environmental factors. (407)

4. List two possible explanations for the higher rate of depression among adolescent girls as compared to boys. (408)
A. ______________________________
B. ______________________________

5. Currently, suicide is the _________ leading cause of death among young people. It has ______________ over the last thirty years. (408)

6. The number of boys who commit suicide exceeds the number of girls by ___________________. Girls make (more/fewer) unsuccessful attempts at suicide than do boys. (408)

7. Describe two types of young people who tend to commit suicide. (408-409)
A.______________________________
____
B.______________________________
____

8. Cite family background factors and precipitating events related to suicide: (409)
Family: ______________________________
Events: ______________________________

9. Teenagers' improved ability to ______________ seems to play a role in the rise in suicide at adolescence. In addition, the ________________________ leads many teenagers to conclude that no one could possibly understand their intense pain. (409)

10. List three ways to help prevent adolescent suicides. (409-410)
A. ______________________________
B. ______________________________
C. ______________________________

11. True or False: Teenage suicides often take place in clusters. (410)

12. Young people under the age of 21 account for ____ percent of police arrests in the United States. (410)

13. The desire for ________________ causes delinquency to rise during early adolescence, remain high in middle adolescence, and then decline into young adulthood. (411)

14. List the two routes to adolescent delinquency and briefly describe the differences between the two. Circle the one that is most likely to lead to a life-course pattern of aggression and criminality. (410-411)

______________________ type: ____

____

______________________ type: ____

____

15. About ________ times as many boys as girls commit major offenses. (411)

16. Describe the family environments of most delinquent youths. (412)

____

17. Describe school and neighborhood environments related to higher rates of delinquency. (412)
School: ____
Neighborhood: ____

18. Cite three general ways to prevent delinquency. (412)
A. ____
B. ____
C. ____

19. Describe characteristics of treatment programs for delinquents that work best. (412)

____

____

ASK YOURSELF . . .

Review the conversation between Louis and Darryl at the opening of this chapter. What identity status best characterizes the two boys? Explain your answer. (see text pages 390, 392)

____

____

____

Summarize personal characteristics associated with identity achievement, and explain how they promote favorable cognitive, emotional, and social development. Why is identity development a lifelong process? (see text pages 392-394)

____

____

____

____

How does Louis and Darryl's lunchtime conversation at the beginning of this chapter reflect the characteristics of adolescent friendship? (see text pages 390, 403-404)

Phyllis likes her 14-year-old daughter Farrah's friends, but she wonders what Farrah gets out of hanging out at Jake's Pizza Parlor with them on Friday and Saturday evenings. Explain to Phyllis what Farrah is learning that will be of value in adulthood. (see text page 404-405)

How might gender intensification contribute to the shallow quality of early adolescent dating relationships? (see text pages 401, 405-406)

Throughout childhood, Mac had difficulty learning, was disobedient, and picked fights with peers. At age 16, he was arrested for burglary. Zeke had been a well-behaved child, but around age 13, he started spending time with the "wrong crowd." At age 16, he was arrested for property damage. Which boy is more likely to become a long-term offender, and why? (see text pages 410-412)

Return to Chapter 11 and reread the sections on adolescent pregnancy and parenthood. What factors do these problems have in common with suicide and delinquency? How would you explain the finding that teenagers who experience one of these difficulties are likely to display others? (see text pages 365-368, 408-412)

SUGGESTED READINGS

Crockett, Lisa J., and Silbereisen, Rainer K. (Eds.). (1999). *Negotiating adolescence in times of social change*. New York: Cambridge University Press. Explores the processes through which societal changes affect the course of adolescent development and adolescents' social and psychological adjustment.

Furman, W., Brown, B. B., & Feiring, C. (Eds.). (1999). *The development of romantic relationships during adolescence*. New York: Cambridge University Press. Examines adolescent romantic relationships and general processes and individual differences within the general context of adolescent development.

Graber, J. A., Brooks-Gunn, J., & Petersen, A. C. (Eds.). (1996). *Transitions through adolescence: Interpersonal domains and contexts.* Mahwah, NJ: Erlbaum. Experts contributing chapters to this volume discuss adolescent transitions in three domains: the peer system, the family system, and school and work contexts. Among the topics considered are friendship, child bearing, school transitions, and low-wage employment. In addition to new research, the authors consider intervention strategies and policy implications.

Gullotta, T. P., Adams, G. R., & Montemayor, R. (Eds.). (1998). *Delinquent violent youth.* Thousand Oaks, CA: Sage. Provides an overview of crime among both urban and rural youths. Issues include how various social factors influence delinquency, treatment for violent behavior, and social policies that prevent crime.

Rest, J., Narvaez, D., Bebeau, M. J., & Thoma, S. J. (1999). *Postconventional moral thinking: A neo-Kohlbergian approach.* Mahwah, NJ: Erlbaum. Discusses problems with Kohlberg's theory and methodology, analyzes recent criticisms, and proposes a new approach called "Neo-Kohlbergian."

8. The desire to obey rules in order to be considered "good" characterizes Stage ______ of Kohlberg's theory. (397-398)
   a. 1
   b. 2
   c. 3
   d. 4

9. When people generate real-life moral dilemmas of their own, moral functioning (398)
   a. becomes more mature.
   b. becomes less mature.
   c. does not change in maturity.
   d. follows no predictable pattern.

10. Research indicates that (399)
    a. the morality of males and females does not differ.
    b. the morality of females tends to stress an ethic of care.
    c. the morality of females is less advanced.
    d. the morality of males is less advanced.

11. Which of the following is NOT associated with advanced moral reasoning? (399-400)
    a. democratic child rearing
    b. years of schooling completed
    c. confrontation and criticism between peers
    d. simpler societal structures

12. During early adolescence, gender intensification is (401)
    a. stronger for boys.
    b. stronger for girls.
    c. the same for girls and boys.
    d. not yet an important issue.

13. Parent-child relations in adolescence can be better understood if we keep in mind that (402)
    a. both parents and teenagers are undergoing a major life transition.
    b. parents are undergoing a major life transition, while teenagers are not.
    c. teenagers are undergoing a major life transition, while parents are not.
    d. neither parents nor teenagers are dealing with major life transitions.

14. At adolescence, sibling interactions (403)
    a. become less intense.
    b. show greater compliance on the part of younger siblings
    c. increase because adolescents invest more time in the relationship.
    d. are rigid.

15. Teenagers in the United States spend more time together outside the classroom than do teenagers in Asian countries. The difference is probably due to (403)
    a. high rates of maternal employment in the United States.
    b. less demanding academic standards in the United States.
    c. differing peer group structures.
    d. both a and b above.

16. ________ is associated with greater intimacy in boy's friendships (404)
   a. Masculinity
   b. Intelligence
   c. Androgyny
   d. Athletic ability

17. Compared to peer groups of middle childhood, adolescent peer groups are (404)
   a. more tightly structured and exclusive.
   b. less tightly structured and exclusive.
   c. more limited to peers of the same sex.
   d. not as constructive in their activities.

18. The achievement of intimacy in dating relationships typically _________ intimacy in same-sex friendships (405)
   a. appears before
   b. lags behind
   c. appears at about the same time as
   d. replaces

19. Peers would be most likely to influence one another's (406)
   a. religious values.
   b. choice of college.
   c. choice of clothes.
   d. career plans.

20. Severe teenage depression (406-407)
   a. occurs more often in girls than boys.
   b. is a normal part of adolescence.
   c. is promoted by environment, not heredity.
   d. requires no treatment.

21. Suicide is (408-409)
   a. not a leading cause of death among young people.
   b. a growing national problem.
   c. on the decline.
   d. higher in adolescence than in old age.

22. Which of the following factors contributes to the sharp rise in suicide from childhood to adolescence? (409)
   a. adolescent impulsiveness
   b. belief in the personal fable
   c. increased emotional distance between parent and child
   d. impersonal school environments

23. When teenagers are asked directly and confidentially, _______ admit that they are guilty of a delinquent offense. (410)
   a. very few
   b. about half
   c. almost all
   d. only boys

24. A major factor related to juvenile delinquency is (411-412)
    a. consistent discipline from parents.
    b. a low-warmth, high-conflict family environment.
    c. teenage exuberance.
    d. social competence.

25. Treating serious juvenile offenders is most effective when (412)
    a. they remain in their own homes and communities, so far as possible.
    b. they are removed from their communities early in treatment.
    c. there is a focus on gaining rapid employment and financial stability.
    d. interventions are brief so as not to interfere with family and occupational functioning.

CHAPTER 13

PHYSICAL AND COGNITIVE DEVELOPMENT IN EARLY ADULTHOOD

BRIEF CHAPTER SUMMARY

Dramatic gains in average life expectancy have occurred over the last century. Both hereditary and environmental factors play a role in life expectancy and physical changes over the lifespan. While some scientists are concerned with increasing human longevity, many more agree that increasing the years of active, vigorous life is an appropriate goal. The programmed effects of specific genes appear to combine with the cumulative effects of random events to contribute to the aging process. The physical changes of aging are very gradual over early adulthood, and the effects are mediated by exercise, nutrition, health practices, and stress levels. Obesity and substance abuse have a major impact on health and longevity.

While they enjoy a wider range of sexual choices than did previous generations, American adults are less sexually active than commonly believed. Heterosexuals and homosexuals, alike choose partners who are similar to themselves and are more satisfied in committed relationships. Sexually transmitted disease, rape, and premenstrual syndrome are important concerns of early adulthood.

Adult cognition changes to reflect an awareness of multiple truths, integration of logic with reality, and tolerance of the gap between real and ideal. Gains in expertise enhance problem-solving as well as creativity, which increasingly involves the formulation of meaningful problems to be solved. Longitudinal findings indicate that intellectual performance improves steadily into middle adulthood and declines only late in life.

College experiences contribute to gains in knowledge and reasoning, revised attitudes and values, enhanced self-knowledge, and career preparation. In societies with many career possibilities, occupational choice is a gradual process. Vocational choices are influenced by personality, parents' occupations, and teachers. Gender-stereotyped messages prevent many women from reaching their career potential. Many young people would benefit from greater access to vocational information. Youth apprenticeships such as those in Germany might improve the limited career options of American high school graduates who do not attend college.

LEARNING OBJECTIVES

After reading this chapter, you should be able to:

13.1. Discuss factors that have contributed to changes in average life expectancy over the twentieth century. (420-421)

13.2. Discuss the extent to which maximum lifespan and active lifespan can and should be improved. (421-423)

13.3. Describe current theories of biological aging, including those at the level of DNA and body cells and those at the level of tissues and organs. (423-425)

13.4. Describe the physical changes of aging, paying special attention to the cardiovascular and respiratory systems, motor performance, the immune system, and reproductive capacity. (425-429)

13.5. Explain the impact of nutrition and exercise on health, and discuss the prevalence, causes, consequences, and treatment of obesity in adulthood. (429-432)

13.6. Name the two most common substance disorders, and discuss the health risks each entails. (432-434)

13.7. Describe sexual attitudes and behavior of young adults today, and discuss factors related to sexually transmitted disease, sexual coercion, and premenstrual syndrome (PMS). (434-439)

13.8. Explain how psychological stress affects health. (Page 439)

13.9. Describe the restructuring of thought in adulthood, drawing on three influential theories. (440-441)

13.10. Discuss the development of expertise and creativity in adulthood. (442)

13.11. Describe general changes in mental abilities assessed on intelligence tests during adulthood. (442-443)

13.12. Describe the impact of a college education on young people's lives, and discuss the problem of dropping out. (444-445)

13.13. Trace the development of vocational choice, and cite factors that influence it. (445-448)

13.14. Discuss career options open to the 25 percent of high school graduates who are not college bound. (448)

STUDY QUESTIONS

Physical Development

Life Expectancy

1. Dramatic gains in ______________________________ provide powerful support for the malleability of biological aging. (421)

2. List the four leading causes of death in young adulthood, in order of prevalence. (421)
A. ______________________________ B. ______________________________
C. ______________________________ C. ______________________________

3. Parental ______________ and ______________ instability in adulthood appear to be strong predictors of age of death. (422-423)

4. Research suggests that individuals who are ______________________ and ______________________ live longer. Describe the personality characteristics of those who may tend to have shorter lives. (422-423)

____

____

5. On average, women live _________ years longer than men. (421)

6. True or False: The United States ranks only nineteenth in average life expectancy among the nations of the world. (421)

7. According to current estimates, the *maximum lifespan*, or the ______________ limit to length of life, varies between ____ and ______ years for most people. (421-422)

8. Gains in average life expectancy are largely the result of reducing ______________ __________ in the first __________ years; expected life for persons age 65 or older has increased (greatly/little). (422)

9. Americans can expect an average *active lifespan*, or years of _____________ life, of about ____ years. This figure has (increased/decreased) in recent years. (423)

Theories of Biological Aging

1. True or False: Unlike a machine, "worn-out" parts of the body usually replace or repair themselves. (424)

2. What two research findings serve as evidence for the existence of "aging genes?" (424)

A. ____

____

B. ____

____

3. According to the "random" view, DNA in body cells is gradually damaged by spontaneous or externally caused _________________. As the damage accumulates, cell _________________________________ is less efficient or abnormal. (424)

4. One probable cause of age-related DNA and cellular abnormalities, implicated in more than 60 disorders of aging, is the release of ___________________________ that form in the presence of oxygen. (424)

5. What foods forestall free radical damage? (424)

____

6. When normally separate fibers cross link, tissue becomes less ___________, leading to many negative consequences at the organ level. What external factors can reduce this effect? (424)

____

7. Gradual failure of the __________________ system, which produces and regulates hormones, is another route to aging. List two examples of decreased hormone production and how they contribute to aging. (424)

A. ____

B. ____

8. Declines in ________________ system functioning, leading to increased susceptibility to disease, seem to be genetically programmed but can be intensified by other aging processes. (424)

Physical Changes

1. Although its functioning under typical conditions does not change, the heart has difficulty delivering enough oxygen to the body during ________________. (425)

2. ____________________________ is a disease in which heavy deposits of fatty plaque collect on the walls of arteries. If present, it begins (early/late) in life. Animal research suggests that ___________________________ may heighten the insult of a high-fat diet. (425)

3. Why has heart disease decreased considerably over the last 10 years? (425)

____

____

4. During physical exertion, respiratory volume (increases/decreases) and breathing rate (increases/decreases) with age. (427)

5. Many studies show that athletic skills reach a peak between ____ and ____ years of age and decline thereafter. (427)

6. True or False: Longitudinal research with master runners indicates that as long as practice continues, performance drops only two percent per decade into the sixties and seventies. (428)

7. ___ cells attack antigens directly, while ___ cells secrete antibodies which capture antigens so they can be destroyed. Receptors on the surface of each of these cells detect (many/one) type of antigen. (428)

8. Partly because of gradual shrinking of the _____________, which promotes maturation and differentiation of T cells, the capacity of the immune system declines after age ________. (428)

9. Stress can also weaken the immune response. Explain why this is so. (428)

____

____

10. Twenty-six percent of women age ________ to ________ are affected by fertility problems, which rise with age. (429)

Health and Fitness

1. What is responsible for SES differences in health over the lifespan? (429)

____

____

2. Currently, ___________ of the American population is obese, and the rates are highest among ______________________________________. (430)

3. List some problems associated with obesity. (430)

____

____

4. List five elements of treatment for obesity that promote lasting behavioral change. (430-431)

A. ____

B. ____

C. ____

D. ____

E. ____

5. List three health problems associated with fat consumption, and indicate which is of primary concern. (431)

A. ______________________________ B. ______________________________

C. ______________________________

6. Cite evidence that excess fat consumption and other societal conditions are largely responsible for the high incidence of heart disease in African Americans. (431)

____

7. List five ways in which exercise helps prevent serious illnesses. (431-432)

A. ____

B. ____

C. ____

D. ____

E. ____

8. How much exercise is recommended for a healthier and longer life? (432)

____

9. Smoking in America has declined (rapidly/slowly) in recent years. (432)

10. List several consequences of smoking. (432-433)

____

____

11. True or False: A variety of long-term treatments are available to help smokers learn skills necessary for avoiding relapse after quitting. (433)

12. At what age does alcoholism usually begin among men and women? (433)

Men: ____________________________ Women: ____________________________

13. Twin and adoption research supports a _______________ role in alcoholism, but personal and environmental conditions contribute, as well. For example, alcohol abuse is (more/less) likely in cultures where it is traditionally used in religious ceremonies and (more/less) likely when it is carefully controlled and a sign of adulthood. (433)

14. List some health problems associated with alcohol use. (433)

____

____

____

15. Describe components of the most successful alcohol treatment programs. (433-434)

____

____

16. Sexual partners usually meet in ______________________________ ways. (434)

17. True or False: Americans have more sexual partners by the age of 50 than they did in the past. However, 71 percent have only one partner in a year's time. (431)

18. Match each of the following theorists or theories with the appropriate explanation for gender differences in sexual attitudes and behaviors. (435)

_____	Emotional intensity of the infant-caregiver relationship is carried over into intimate ties for girls, but for boys it is disrupted by formation of a "masculine" gender role stressing independence	1. Chodorow and Gilligan
_____	Women receive more disapproval for having numerous sexual partners, while men are sometimes rewarded with admiration and social class	2. Ethological theory
_____	Since sperm are plentiful and ova far less numerous, women must be much more careful in their selection of partners	3. Social learning theory

19. True or False: Young women's complaints that many men are not interested in long-term commitments are generally unfounded. (435)

20. Sexual activity (increases/decreases) through the twenties as people either marry or cohabitate and declines around age _____ due to the demands of daily life. (434)

21. True or False: As number of sexual partners increases, satisfaction with the sex life increases. (435)

22. List the two sexual problems most frequently reported by women and the two most frequently reported by men. (435)

Women: ____

Men: ____

23. (Men/Women) judge homosexuals more harshly. (436)

24. True or False: Like heterosexual couples, homosexuals appear to choose partners similar to themselves, have sex more often and are more satisfied in the context of committed relationships, and engage in modest levels of sexual activity. (436)

25. One in ____ Americans are likely to contract a sexually transmitted disease sometime in their lives, and it is at least twice as easy for a (man/woman) to be infected. (432)

26. List ways in which AIDS can be contained and reduced. (436)

____

____

27. One study reported that ________ of American women have experienced sexual coercion. Nearly ____ percent of men admitted to obtaining sex through force. (437)

28. Briefly describe the personal characteristics of men who engage in sexual assault. (437)

____

____

____

____

29. List two cultural forces that contribute to sexual coercion. (437)

A. ______________________ B. ______________________

30. Summarize the consequences of rape and ongoing sexual assault. (437-438)

____

____

____

____

31. List four critical features of treatment for rape victims. (438)

A. ____

B. ____

C. ____

D. ____

32. About ____ women worldwide experience some form of PMS, but only ____ percent of women experience symptoms severe enough to interfere with academic, occupational, or social functioning. (438)

33. True or False: Helping people who are isolated develop and maintain satisfying interpersonal relationships is not as important as the other health interventions mentioned. (439)

Cognitive Development

Changes in the Structure of Thought

1. ____________ found that younger college students tend to engage in__________________ thinking, in which information and values are divided into right and wrong, good and bad. (440)

2. Older college students are more likely to use ________________ thinking, in which the possibility of absolute truth is given up in favor of multiple truths which are relative to their context. (440)

3. According to Schaie, the goals of mental activity shift from ________________ knowledge to ________________ during adulthood to achieve goals, maintain social obligations, and remain consistent with values and attitudes. List the four stages of his model and the age period with which each is associated. (440-441)
A. ________________________________
B. ________________________________
C. ________________________________
D. ________________________________

4. In Labouvie-Vief's view, the movement from ________________ to ________________ thought is a structural advance in which logic becomes a tool to solve real-world problems. (441)

5. In the course of balancing various roles, adults give up their need to resolve ________________ and accept them as a part of life. (441)

6. Summarize the qualitative transformations in adult thinking presented in the three theories just discussed. (441)
A. ________________ B. ________________
C. ________________________________

Information Processing: Expertise and Creativity

1. Compared to novices, experts ________________ and ________________ more quickly and effectively. (442)

2. How do the creative products of adults differ from those of children? (442)

____

3. Creative accomplishment typically rises in ________________, peaks around ________________ and then gradually declines. However, exceptions to this trend indicate that it is more a function of "________________ age" than actual age. (442)

4. List four qualities necessary for creativity in addition to expertise and situational factors. (442)
A. ________________________________
B. ________________________________
C. ________________________________
D. ________________________________

Changes in Mental Abilities

1. Schaie's longitudinal-sequential study revealed that cross-sectional declines in IQ scores were actually due to ________________, such as improvements in education and health. Intellectual performance actually improves from ________________ to ________________ adulthood with a modest drop that does not occur until late in life. (443)

The College Experience

1. List four psychological changes, in addition to increased knowledge, that occur during the college years. (444)
A. ____________________
B. ____________________
C. ____________________
D. ____________________

2. By what two factors is the impact of college jointly determined? (444)
A. ____________________
B. ____________________

3. List four student characteristics that may make adaptation to college particularly difficult. (444)
A. ____________________ B. ____________________
C. ____________________ D. ____________________

4. Factors involved in the decision to drop out of college are usually (typical/ atypical) problems of early adulthood. (445)

Vocational Choice

1. What guides career preferences during the fantasy period of vocational development, typically ranging from early to middle ____________________? (445)
____

2. During the tentative period, typically occurring between ages ___ and ___, individuals evaluate vocational options first in terms of their ________________ and later take account of their ____________________. (445)

3. In the realistic period, occurring during ____________________ and ____________________, young people narrow their options through further ________________ and enter a final phase of ____________________ in which they focus on a general vocational category. (445)

4. Match each of the following personality types that affect vocational choice with the appropriate description. (445-446)

____	Likes well-structured tasks and values social status; tends to choose business occupations	1. Investigative
____	Prefers real-world problems and work with objects; tends toward mechanical occupations	2. Social
____	Adventurous, persuasive, and a strong leader; drawn toward sales and supervisory positions	3. Realistic
____	Enjoys working with ideas; drawn toward scientific occupations	4. Artistic
____	Has a high need for emotional and individual expression; drawn to artistic fields	5. Conventional
____	Likes interacting with people; drawn toward human services	6. Enterprising

5. The relationship between personality and vocational choice is (weak/moderate/strong). (446)

6. Identify two reasons, in addition to years of educational attainment, why vocational choice is typically similar to that of one's parents. (446)
A.____
____
B.____
____

7. True or False: When asked who had the greatest influence on their choice of a field of study, college freshmen most often mentioned high school teachers. (446)

8. Identify two influences on women's growing attraction to nontraditional careers. (446)
A. ______________________ B. ______________________

9. A decline in estimates of their ______________________, a shift in their expectations toward ______________________, and concerns about combining work with ______________________ are all gender-stereotyped issues which contribute to women's lower occupational attainment. (447)

10. Identify four common experiences of women who achieve high career aspirations. (447)
A. ____
B. ____
C. ____
D. ____

11. What type of students are at risk for becoming "drifting dreamers"? (447)
____
____

12. How can high school and colleges help high-ambition/low-knowledge students choose an occupational goal and complete their educational plans? (448)
____
____
____

13. High school graduates who do not attend college have (more/fewer) work opportunities than they did several decades ago. More than ______________ of them are unemployed. (448)

14. In Germany, apprenticeships are jointly planned and supported by ______________ and ______________. As a result of successful training programs, German young people are established in secure, well-paid jobs with advancement opportunities by the ages of ______________. (449)

15. Identify three major challenges to the implementation of successful apprenticeship programs in the United States. (449)
A. ____
B. ____
C. ____

ASK YOURSELF . . .

Explain why the United States ranks only nineteenth in life expectancy among the world's nations. (For help in answering this question, return to the section on public policies and lifespan development in Chapter 2, pages 68-70.) (see text page 421)

Len noticed the term *free radicals* in an article on adult health in his local newspaper, but he doesn't understand how they contribute to biological aging. Explain the role of free radicals to Len. (see text page 425)

Penny is a long-distance runner for her college's track team. She wonders what her running performance will be like 10 or 20 years from now. Describe physical changes and environmental factors that will affect Penny's long-term athletic skill. (see text pages 427-428)

List as many factors as you can that may have contributed to heart attacks and early death among Sharese's relatives. (see text page 425)

Tom began going to a health club three days a week after work. Soon the pressures of his job and his desire to spend more time with his girlfriend convinced him that he no longer had time for so much exercise. Explain to Tom why he should keep up his exercise regimen, and suggest ways to fit it into his busy life. (see text pages 431-432)

Why are people in committed relationships likely to be more sexually active and satisfied than those who are dating several partners? (see text pages 434-436)

While taking a college human development course, Marcia wrote a paper on disciplining children. In the paper, she explained that behaviorists and psychoanalysts would view physical punishment as ineffective, but for different reasons. Both viewpoints, she concluded, could be valid. Explain how Marcia's reasoning illustrates postformal thought. (see text page 440)

As a college sophomore, Steven took a course in decision making but didn't find it very interesting. At age 25, newly married and employed, he realized the value of what the professor had taught him. How do changes in the structure of thought help explain Steven's revised attitude? (see text pages 440-441)

____

____

____

How does the development of creativity in adulthood illustrate assumptions of the lifespan perspective? (Return to Chapter 1, pages 7-12 and 442)

____

____

____

During her freshman year of college, Sharese participated in a program in which a professor and five students met once a month to talk about common themes in three courses that the students took together. Why is this program likely to increase student retention? (see text pages 444-445)

____

____

____

Jordan, a community college student, wants to become a teacher but has little idea of the course work needed to achieve his vocational goal. Why might the gap between Jordan's ambition and knowledge of his chosen vocation have negative consequences for his educational and occupational attainment? What steps can colleges take to prevent students like Jordan from becoming "drifting dreamers?" (see text pages 447-448)

____

____

____

____

SUGGESTED READINGS

Medina, J. J. (1996). *The clock of ages: Why we age—how we age—winding back the clock.* Cambridge: Cambridge University Press. Examines the processes of aging and death. First, aging and death are discussed from an evolutionary context. Then, the author describes aging in terms of changes in the tissues and organs of the human body. The final section covers theories of aging and factors that play a role in increasing longevity.

U. S. Department of Health and Human Services. (2000). *Healthy people 2010: Understanding and improving health.* Washington, DC: U.S. Government Printing Office. Outlines a comprehensive, nationwide health promotion and disease prevention agenda. It is designed to serve as a roadmap for improving the health of all people in the United States during the first decade of the twenty-first century.

Ward, T. B., Smith, S. M., & Vaid, J. (1997). *Creative thought: An investigation of conceptual structures and processes.* Washington, DC: American Psychological Association. Examines the nature of creative thought, emphasizing the generative aspects of creativity or how old concepts are used to bring about new ideas.

Whitbourne, S. K. (1996). *The aging individual: Physical and psychological perspectives.* New York: Springer. Promotes understanding of how biology and psychology interact in the aging individual. The cardiovascular and respiratory systems, the nervous system, sensation and perception, mobility, intellectual functioning, identity, and personality are among the topics considered. Themes include the interplay between physical and cognitive change and ways individuals think about their own aging.

PUZZLE 13.1 TERM REVIEW

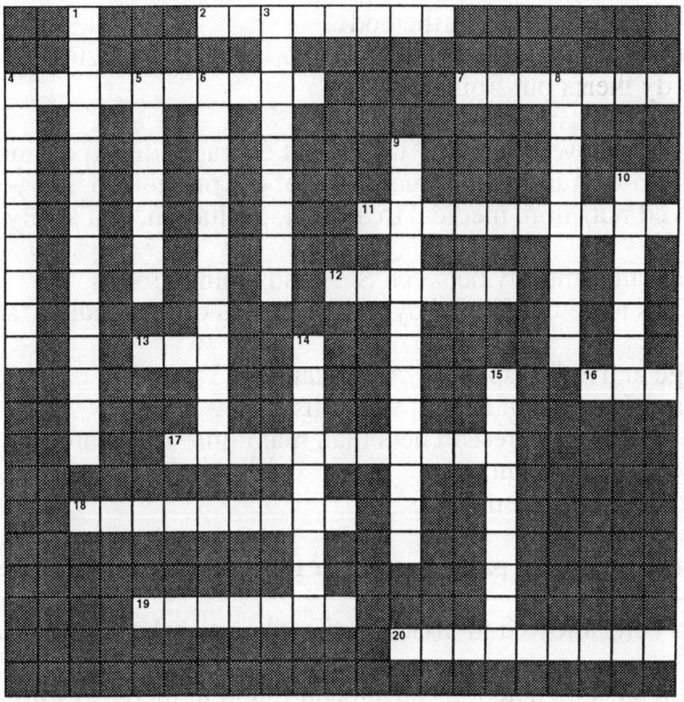

Across

2 Free _____ form in the presence of oxygen and destroy cellular material.

4 _____ thought: cognitive development beyond Piaget's formal operation

7 _____ life expectancy: number of years an individual born in a particular year can expect to live

11 _____ aging: genetically programmed declines in organ functioning universal in all members of each sex within a species

12 _____ lifespan: years of vigorous, healthy life

13 Physical and psychological symptoms that appear 6 to 10 days prior to menstruation (abbr.)

16 Amount of energy6 the body uses at complete rest (abbr.)

17 Schaie's stage in which the goal of mental activity is knowledge acquisition

18 Period of vocational development in which adolescents weigh options against interests, abilities, and values

19 Cross-_____ theory asserts that formation of bonds between normally separate fibers causes connective tissues to become less elastic.

20 Acquisition of extensive knowledge in a field or endeavor

Down

1 Schaie's stage in which cognitive skills are adapted to meet expanding responsibilities to others

3 _____ thinking divides information, values, and authority into right and wrong, good and bad, we and they.

4 Thought in which logic becomes a tool to solve real-world problems

5 Period of vocational development involving career exploration through make-believe play

6 Thinking of older college students, who favor multiple truths relative to context

8 Period of vocational development involving a focus on a career category and the choice of a single career

9 Schaie's stage in which interests, attitudes, and values are re-examined and used as a guide for knowledge acquisition and application

10 _____ lifespan: genetic limit to length of life for a person free of external risk factors

14 Schaie's stage in which cognitive skills are adapted to long-term goal achievement

15 _____ stage: more advanced form of the responsibility stage characterized by people at the helm of large organizations

SELF-TEST

1. Which of the following is true of biological aging? (420-421)
   a. Like physical growth, it is an asynchronous process.
   b. It affects all structures of the body.
   c. Individual differences are small due to its genetic origin.
   d. The body wears out from use.

2. Which of the following is NOT true about average life expectancy? (421)
   a. Twentieth-century gains equal those of the previous 5,000 years.
   b. Improved nutrition, medical treatment, sanitation, and safety are responsible for gains.
   c. It varies substantially between SES and ethnic groups.
   d. It appears to be unaffected by marital status or personality factors.

3. The average active life span for Americans (423)
   a. is around 50 years of active, vigorous life.
   b. is less important to researchers than maximum life span.
   c. has declined in recent years.
   d. has increased in recent years.

4. Which of the following provide support for the "programmed" effects of "aging genes?" (424)
   a. Human cells allowed to divide in the laboratory have a lifespan of about 50 divisions.
   b. DNA breaks and deletions have been found to increase with age in a variety of animal species.
   c. Free radicals have been found to damage cellular material.
   d. Longevity does not appear to be a family trait.

5. Cross-linkage theory suggests that (424)
   a. longevity is fostered by supportive tissue links between organs.
   b. bonds between protein fibers in the connective tissue contribute to aging.
   c. the cross-linking of fibers is unaffected by external factors.
   d. both b and c above are true.

6. Which of the following is NOT true of age-related changes in cardiovascular functioning? (425)
   a. In healthy individuals, the heart's ability to meet oxygen needs under typical conditions does not change in adulthood.
   b. Changes in functioning are due to a decrease in maximum heart rate and greater rigidity of the heart muscle.
   c. If present, atherosclerosis usually begins late in life.
   d. Heart disease has decreased considerably since mid-century.

7. The capacity of the immune system to offer protection against disease (428)
   a. decreases through adolescence and inclines after age 20.
   b. is influenced by changes in the thymus.
   c. is undermined by stress hormones.
   d. is both b and c.

8. Which of the following is NOT consistently related to nearly all indicators of health? (429)
   a. income
   b. education
   c. occupational status
   d. marital status.

9. What underlies the rising rates of obesity in industrialized nations? (430)
   a. Heredity
   b. A reduction in basal metabolism rate
   c. Reduced physical labor and increased fat intake
   d. Increased alcohol consumption

10. Which of the following is NOT true of cigarette smoking? (432-433)
   a. Smoking has declined only very slowly in recent years.
   b. The cumulative effects of smoking include limited night vision, male sexual impotence, and wrinkling of the skin.
   c. After quitting for a full year, about 30 percent start smoking again.
   d. The benefits of quitting include a return to nonsmoker disease risk levels within 3 to 8 years.

11. Alcoholism (433)
   a. is almost exclusively related to genetic factors.
   b. is more likely where alcohol is carefully controlled.
   c. is most effectively treated in a hospital setting.
   d. usually has an earlier onset for women than for men.

12. _____ of Americans age 18 to 59 have intercourse as often as twice a week and sexual activity is highest for _______ couples. (434)
   a. One-third; married/cohabiting
   b. Half; married/cohabiting
   c. One-third; dating
   d. Half; dating

13. Among what group is the rapid spread of AIDS now most likely? (436)
   a. women
   b. children
   c. gay men
   d. intravenous drug users and their partners

14. Which of the following is NOT true of men who engage in sexual assault? (437)
   a. They tend to believe in traditional gender roles.
   b. They tend to interpret women's social behavior incorrectly.
   c. They tend to be less educated and of lower socioeconomic status.
   d. They tend to approve of violence against women.

15. Which of the following is NOT one of the ways in which stress affects health? (439)
   a. It elevates blood pressure.
   b. It interferes with immune functioning.
   c. It increases digestive activity.
   d. It is linked to hypertension.

16. William Perry's theory suggests that young adults make a transition from dividing information into absolutes, such as right and wrong, to (440)
    a. dualistic thinking.
    b. relativistic thinking.
    c. adaptive cognition.
    d. multiplistic thinking.

17. Which of Schaie's stages occurs in early adulthood? (440-441)
    a. acquisitive stage
    b. responsibility stage
    c. achieving stage
    d. executive stage

18. Which theorist introduced the idea that thought moves from being hypothetical to being pragmatic in early adulthood? (441)
    a. Perry
    b. Schaie
    c. Labouvie-Vief
    d. Piaget

19. Although there are important exceptions, people are likely to be the most creative at which of the following age periods? (442)
    a. teens and early twenties
    b. twenties and thirties
    c. thirties and forties
    d. forties and fifties

20. Mental abilities in adulthood as assessed by intelligence tests (443)
    a. continue to increase into the fifties and sixties.
    b. drop off sharply in the middle thirties.
    c. depend on cohort effect.
    d. gradually decline once individuals' formal education ends.

21. While 75 percent of high school graduates enroll in institutions of higher learning, _____ percent drop out during their freshman year. (444)
    a. 40
    b. 50
    c. 60
    d. 70

22. Which period of vocational choice is experienced during early and middle adolescence? (445)
    a. fantasy period
    b. tentative period
    c. realistic period
    d. exploratory period

23. Which of the following is NOT one of the primary factors that influence vocational choice, as discussed in the text? (445-447)
    a. personality
    b. family influences
    c. teachers
    d. pay and benefits

24. Women get higher grades than men in secondary school and female valedictorians outperform their male counterparts in college. Once they enter the workforce, women (447)
    a. achieve at higher levels than do men.
    b. are confident in their abilities to excel in the professional arena.
    c. achieve at lower levels than men and take traditional career paths.
    d. receive more social support for achievements than do men.

25. Which is NOT a traditionally feminine profession? (447)
    a. nursing
    b. education
    c. literature
    d. veterinary science

CHAPTER 14

EMOTIONAL AND SOCIAL DEVELOPMENT IN EARLY ADULTHOOD

BRIEF CHAPTER SUMMARY

According to Erikson, young adults face the conflict of intimacy versus isolation. Successful resolution requires a balance of independence and intimacy. Levinson suggested that the life course consists of a series of eras in which individuals revise their life structure to meet changing needs. Vaillant expanded Erikson's stages to include the pursuit of intimacy and career during the twenties and thirties. Conformity to a culturally determined timetable for major life events gives young adults confidence, while departure from it can lead to distress.

As in other age periods, intimate partners in early adulthood resemble each other in characteristics and background. The ethological and social learning theories provide perspectives on how gender roles influence criteria for mate selection. Romantic love is the basis for marriage in Western cultures, but the balance among passion, intimacy, and commitment changes as relationships progress. Friendships continue to be based on trust with women's friendships remaining more intimate, and sibling relationships often become stronger. Other-sex friendships occur less often and do not last as long as same-sex friendships.

Significant diversity characterizes the sequence and timing of the modern family life cycle. While marriages are becoming more egalitarian, the majority are still relatively traditional in terms of gender roles. Modern couples are having fewer children and postponing parenthood longer than in past generations. The transition to parenthood profoundly alters the lives of husbands and wives and can threaten marital satisfaction. Special interventions can ease this transition.

More adults are remaining single today and cohabitation has increased dramatically. The number of couples who choose to remain childless has increased, as well. About half of marriages will end in divorce, and many divorcees will remarry, resulting in blended families with unique challenges. Never-married parenthood has increased and often contributes to financial difficulties. Families headed by homosexuals fare well except for difficulties related to living in an unsupportive society.

Men's career paths are generally continuous, while women's are discontinuous due to child bearing. Although women and ethnic minorities have entered nearly all professions, they tend to be concentrated in less well-paid positions and face unique challenges. Dual-earner couples, now the norm, must resolve more complex career decisions than single-earner families, but benefit from increased income and self-fulfillment of the wife.

LEARNING OBJECTIVES

After reading this chapter, you should be able to:

14.1. Describe Erikson's stage of intimacy versus isolation and related research findings. (456-457)

14.2. Describe and evaluate Levinson's and Vaillant's psychosocial theories of adult personality development, noting how they apply to both men's and women's lives. (457-450)

14.3. Explain how the social clock can affect personality in adulthood. (459-460)

14.4. Explain the role of romantic love in young adults' quest for intimacy, and indicate how the balance among its components changes as a relationship develops. (460-463)

14.5. Describe early adulthood friendships and sibling relationships. (464-465)

14.6. Cite factors that influence loneliness, and explain its role in early adult development. (465-466)

14.7. Trace phases of the family life cycle that are prominent in early adulthood, and cite factors that influence these phases. (466-475)

14.8. Discuss the diversity of adult lifestyles, paying special attention to singlehood, cohabitation, and childlessness. (475-477)

14.9. Discuss today's high rates of divorce and remarriage, and cite factors that contribute to them. (477-478)

14.10. Discuss challenges associated with variant styles of parenthood, including remarried parents, never-married parents, and gay and lesbian parents. (478-480)

14.11. Discuss men's and women's patterns of vocational development, and cite challenges faced by women and ethnic minorities and couples seeking to combine careers and family life. (480-484)

STUDY QUESTIONS

Erikson's Theory: Intimacy Versus Isolation

1. True or False: All modern theories have been influenced by Erikson's vision of adult personality development. (456)

2. The critical psychological conflict of young adulthood is ________________ versus ________________, reflected in the young person's thoughts and feelings about making a permanent commitment to an intimate partner. (456)

3. Describe the characteristics of individuals who have achieved a sense of intimacy and those who have a sense of isolation. (456-457)

Intimacy : ____

____

Isolation: ____

____

4. According to Erikson, successful resolution of intimacy versus isolation prepares the individual for ________________________. (457)

Other Theories of Adult Psychosocial Development

1. In Levinson's theory, each era, or stage, begins with a ________________ lasting about 5 years which is followed by a stable period in which individuals concentrate on building a ____________________ that harmonizes inner-personal and outer-social demands. (457)

2. Describe differences in the life dreams of men and women. (458)
Men: ____
____
Women: ____
____

3. Young adults generally form a relationship with a ______________ who facilitates the realization of their dream. (458)

4. During the age 30 transition, young people reevaluate their __________________ and try to change aspects they find inadequate. Those who stressed ______________ often become more concerned with ______________, and vice versa. (458)

5. For men, Levinson describes the period from age 33 to 40 as one of ______________ ______________. They try to establish a stable niche in society by anchoring themselves more firmly in ______________________, ________________________, and ______________________. (459)

6. Not until ________________________ do many women attain the professional and community-related stability typical of men in their thirties. (459)

7. Briefly summarize typical concerns of men during the following general age periods as identified by Vaillant. (459)
Twenties: ________________________ Thirties: ________________________
Forties: ________________________ Fifties: ________________________

8. Identify three limitations of Levinson's and Vaillant's theories. (459)
A. ____
B. ____
C. ____

9. Being on-time or off-time in development can profoundly affect ________________, since adults measure their progress through social comparison. (460)

10. Describe the characteristics of women born in the 1930s who followed "feminine" and "masculine" social clocks and those who did not follow either. (460)
Feminine: ____
Masculine: ____
Neither: ____

Close Relationships

1. True or False: In selecting a mate, research suggests that "opposites attract." (461)

2. Men and women differ in the importance they place on certain characteristics when selecting a mate. How do the ethological and social learning perspectives explain this difference? (461)
Ethological:____

____

Social learning:____

____

3. Recollections of childhood ____________________ patterns are strong predictors of ____________________ relationships in adulthood. (462-463)

4. List the three components of love presented in your text. (461)
A. ____________________ B. ____________________
C. ____________________

5. In Eastern countries, ____________________ is recognized and viewed positively throughout life, while in the United States it is usually regarded as ____________________. (463)

6. Asians are more likely to stress ____________________ and ____________________ considerations in mate selection rather than physical attraction and emotional intensity. (463)

7. Cite three benefits of adult friendship. (464)
A. ____________________ B. ____________________
C. ____________________

8. Sharing thoughts and feelings may occur (more/less) in friendships than in marriage. Commitment in friendship is (more/less) strong over the life course. (464)

9. When together, female friends say they prefer to ____________________ while male friends prefer to ____________________. (464)

10. Androgynous individuals disclose (more/less) intimate information than do traditional individuals. Married men disclose more towards their (same-sex friends/spouses). (464)

11. Other-sex friendships increase with age for (men/women) and (men/women) are more likely to feel sexually attracted to an other-sex friend. (464-465)

12. As young people marry, siblings become (more/less) frequent companions than they were in adolescence. (465)

13. In Vaillant's study of well-educated men, what was the best predictor of emotional health at age 65? (465)

____

14. Loneliness is at its peak during the ____________________ and ____________________ after which it declines steadily into the seventies. (465)

15. Loneliness is especially intense after ______________________________.
When romantically uninvolved, (men/women) feel lonelier; when married, (men/women) feel lonelier. (465)

16. Describe personal characteristics associated with persistent loneliness. (466)

____
____
____

17. Describe the potential benefits of occasional, mild loneliness. (466)

____
____

The Family Life Cycle

1. The average age of leaving the family home has (increased/decreased) in recent years. (466)

2. Nearly ________ of young adults return home for a brief time after initial leaving. Those who departed to marry are (most/least) likely to return. Those who left because of family conflict (usually/rarely) return. (466-467)

3. Economically advantaged young people are (more/less) likely to live independently before marriage. Members of ethnic minorities are (more/less) likely to leave home before marriage. (467)

4. List at least one advantage of leaving home when prepared for independence and at least one potential disadvantage of early departure. (467)
A.____
B.____
____

5. Nearly ____ percent of Americans marry at least once. (467)

6. When backgrounds are ______________________, married couples face more challenges in achieving a successful transition to married life. (468)

7. Evidence suggests that it is better to marry (earlier/later). Explain why this is so. (468)

____
____

8. True or False: Even when women are employed, they average almost three times as many hours of housework as their husbands. (468)

9. (Men/Women) are much more likely to report being happily married. Relationship quality has a greater impact on mental health for (men/women). (469)

10. Crime studies report higher rates of assaulting (men/women) and physically injured (men/women). List three reasons both husbands and wives give for abusing their spouse. (470)
A.______________________________ B.______________________________
C.______________________________

11. Describe factors that contribute to spouse abuse. (470-471)
Psychological:____

Family: ____

Cultural: ____

12. List five reasons why many people do not leave destructive relationships before abuse escalates. (471)
A. ____
B. ____
C. ____
D. ____
E. ____

13. Of the small number of batterers who agree to participate in treatment, at least ________ continue their violent behavior with the same or a new partner. (471)

14. List three components in which marital happiness is grounded. (469)
A. ____
B. ____
C. ____

15. In a study of college students, more men said that their partner should be (inferior/superior) to themselves and more women said that their partner should be (inferior/superior). Under these conditions, women are likely to __________________ their abilities. (469-470)

16. True or False: Most couples spend little time reflecting on the decision to marry before their wedding day. (471)

17. Women who work in __________________ positions are less likely to become parents than women in ________________________ careers. (472)

18. List the two reasons for having children that are most important to American parents across ethnic groups and geographic regions. (472)
A. ____
B. ____

19. List the two disadvantages of parenthood mentioned most often by young adults. (472)
A. ______________________________ B. ______________________________

20. Entry of children into the family usually causes the roles of husband and wife to become more _____________________, a transition that is most difficult for young mothers who have been _______________________________. (472)

21. List two ways in which waiting until the late twenties or thirties to bear children eases the transition to parenthood. (473)
A. ____
B. ____

22. Describe benefits of fathers' involvement with child rearing. (473)

____

____

23. Describe interventions that ease the transition to parenthood for parents not at risk for problems. How should interventions differ for high-risk parents? (473)

A.____

____

B.____

____

24. Briefly describe benefits of maintaining a small family. (473)

____

____

25. Identify some benefits of child rearing for adult development. (474)

____

____

26. As adolescents move from childhood to adulthood, they alternately become ____________________ and experiment with ____________________, making them appear unpredictable to parents. (474)

27. List three advantages of parent education programs. (474-475)

A. ____

B. ____

C. ____

The Diversity of Adult Lifestyles

1. Recent trends suggest that (few/many/most) Americans will spend a substantial part of their adult lives single, and about __________ percent will stay that way. (475)

2. Explain why women in prestigious careers and men in blue collar careers are overrepresented among singles after age 30. (475)

____

____

3. List the two most often mentioned advantages of singlehood and five drawbacks of singlehood. (476)

Two Advantages: ____

Five drawbacks: ____

____

4. Explain why the following age periods are stressful for many singles. (476)

Late 20s: ____

Mid 30s (for women):____

5. ________ times as many American couples are cohabiting in the 1990s as did in 1970. About ______________ of these households include children. (476)

6. For some, cohabitation serves as ________________ for marriage, while for others it is an __________________ to marriage. (476)

7. American cohabitors typically marry to confirm their ____________________ ____________________, while their Western European counterparts view these sentiments as attached to cohabitation itself. (476)

8. Describe ways in which American cohabitors differ from married or separate-residence couples. (476)

____
____
____
____

9. True or False: In contrast to heterosexual cohabitors, homosexual couples report strong commitment. (477)

10. True or False: Cohabiting is an effective way to avoid bitter legal conflicts in the event of separation. (477)

11. Describe the typical characteristics of individuals who are voluntarily childless. (477)

____
____
____

12. Childlessness interferes with adjustment and life satisfaction only when it is ____. (477)

13. Identify two styles of interaction found in many couples who split up. (478)

A. ____
B. ____

14. List four factors that increase a couple's chance of divorce. (478)

A.______________________________ B.______________________________
C.______________________________ D.______________________________

15. Identify two societal trends related to the current high divorce rate. (478)

A. ______________________________ B. ______________________________

16. Briefly describe the short-term consequences of divorce for men and women. (478)

____
____

17. True or False: Most divorced women prefer their new lifestyle to an unhappy marriage. (478)

18. List four reasons why many remarriages break up. (478)

A. ____
B. ____
C. ____
D. ____

19. Stepparents often view the parent as too (harsh/lenient), and vice versa, differences that can become major issues that divide remarried couples. (479)

20. (Stepmothers/Stepfathers), especially, are likely to experience conflict and poor adjustment. (479)

21. Stepfathers (with/without) children of their own have an easier adjustment. (479)

22. African-American women postpone childbirth (more/less) and marriage (more/less) than do all other American ethnic groups. (479)

23. Black mothers rely heavily on ________________________________ to help with the care of their children. When they do marry, they (do/do not) report the relationship problems typical of blended families. (479)

24. Children of gay and lesbian parents are (more/equally/less) well adjusted when compared to children of heterosexual parents. (480)

25. When children were ______________ or conceived through ______________ ____________________, partners of homosexual parents tend to be more involved than when children originated in a previous heterosexual relationship. (480)

26. What is the greatest concern of gay and lesbian parents regarding their children? (480)
____

Vocational Development

1. While ________ of young people aspire to professional occupations, only ______ percent of the work force attains them. (481)

2. As they become aware of the gap between expectations and reality, people in their twenties change jobs every _____ years, on average. (481)

3. Advancing little in career predicts early __________________ from work, while becoming very successful is associated with __________________ job involvement over time. (481)

4. Young people who are ____ tend to set career aspirations that are too high or too low and achieve less than their abilities would permit. (481)

5. Success in a career also often depends on the formation of a ______________ relationship. (481)

6. For every dollar earned by a man, the average woman earns only ___ cents. (481)

7. For women, career planning is often __________________ and subject to considerable ______________. (481)

8. What three factors slow women's advancement even when they do enter high-status professions? (481-482)
A. ____
B. ____
C. ____

9. What three common themes in women's occupational life stories emerged from interviews with highly successful American women? (482-483)
A.____
B.____
C.____

10. In the "Termite" study, which women were most and least satisfied with their lives at age 60? (482)
Most satisfied: ____
Least satisfied: ____

11. The dominant family form today is the ________________ marriage. (482)

12. Work-family role conflict is greatest for ______________ and is especially intense when they are in low-status occupations with ____________ schedules and little worker ________________. (483)

13. Having a dual-earner family makes _____________ decisions more complex and difficult. (483-484)

ASK YOURSELF . . .

Return to Chapter 1 and review commonly used methods of studying human development. Which method did Vaillant use to chart the course of adult life? What are its strengths and limitations? (see text pages 28-31, 459)

____
____
____

In Levinson's theory, during which periods are people likely to be moving within a social clock timetable? When are they likely to be reflecting on and reconsidering this schedule? (see text pages 457-460)

____
____
____

Using the concept of the social clock, explain why Sharese was so conflicted about getting married to Ernie after she finished graduate school. (see text page 460)

____
____
____

After dating for 2 years, Mindy and Graham reported that their love and satisfaction with their relationship were greater than during the first few months they had known each other. What features of communication probably contributed to a deepening of their bond, and why is it likely to endure? (see text pages 460-462)

____
____
____

Claire and Tom, both married to other partners, got to know each other at work and occasionally have lunch together. What is each likely to gain from this other-sex friendship? (see text pages 464-565)

After her wedding, Sharese was convinced she had made a mistake and never should have gotten married. Cite factors that helped sustain her marriage to Ernie and that led it to become especially happy. (see text pages 467-469)

Suggest several ways in which a new mother and father can help each other make an effective adjustment to parenthood. (see text pages 472-473)

Return to Chapter 10, pages 334-338 and review the impact of divorce and remarriage on children and adolescents. How do those findings resemble outcomes for adults? What might account for the similarities? (see text pages 477-478)

After dating for a year, Wanda and Scott decided to live together. Wanda's mother heard that couples who cohabit before marriage are more likely to get divorced. She worried that cohabitation would reduce Wanda and Scott's chances for a successful life together. Is her fear justified? Why or why not? (see text pages 476-477)

Heather climbed the career ladder of her company quickly, reaching a top-level executive position by her early thirties. In contrast, Sharese and Christy did not attain managerial roles in early adulthood. What accounts for the disparity in career progress of the three women? (see text page 481-482)

Work life and family life are inseparably intertwined. Explain how this is so in early adulthood. (see text pages 482-484)

SUGGESTED READINGS

Cowan, C. P., & Cowan, P. A. (1999). *When partners become parents: The big life change for couples* (2nd ed.). Mahwah, NJ: Erlbaum. Explores the joys, challenges, and changes that couples face with the transition to parenthood.

Hetherington, E. M. (Ed.). (1999). *Coping with divorce, single parenting, and remarriage: A risk and resiliency perspective*. Mahwah, NJ: Erlbaum. Explores individual, familial, and extrafamilial risk and protective factors in an attempt to explain parents' and children's responses to events associated marriage, divorce, life in a single parent household, and remarriage.

Molfese, V. J., & Molfese, D. L. (Eds.). (2000). *Temperament and personality development across the lifespan*. Mahwah, NJ: Erlbaum. Provides an overview of historical issues, theory, and research important for the understanding of temperament in infancy and childhood, and personality in adolescence and adulthood.

PUZZLE 14.1 TERM REVIEW

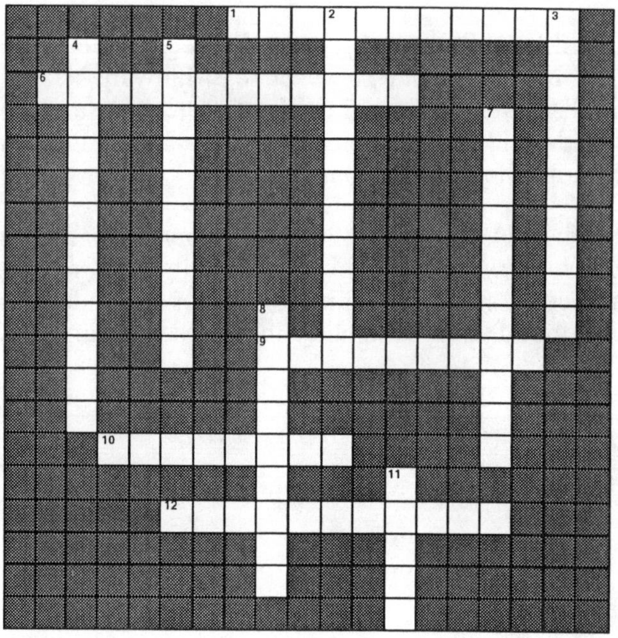

Across

1 _____ marriage involves clear division of husband's and wife's roles, with the man as head of the household and economic provider, and the woman devoting herself to care and nuturance of husband, children, and home.

6 Love based on warm, trusting affection and caregiving

9 Life _____: In Levinson's theory, the underlying pattern of a person's life at a given time, consisting of relationships with significant others which are recognized during each period of adult development

10 Positive outcome of Erikson's conflict of young adulthood

12 _____ _____ cycle: a sequence of phases that characterizes the development of most families around the world (2 words)

Down

2 ____ - ___ marriage: family in which both husband and wife are employed (2 words)

3 Feeling of unhappiness resulting from a gap between actual and desired social relationships

4 Lifestyle of unmarried couples who have an intimate, sexual relationship and share a residence

5 Love based on intense sexual attraction

7 _____ marriage involves sharing of power and authority between husband and wife, with both trying to balance time and energy between work, children, and spouse.

8 Negative outcome of Erikson's conflict of young adulthood

11 Social _____: age-graded expectations for life events

SELF-TEST

1. Which is NOT evidence of a secure sense of intimacy? (456-457)
   a. cooperative behavior
   b. tolerance and acceptance of others, despite differences
   c. comfortable being alone and with others
   d. a competitive nature

2. Which is NOT true? (457)
   a. Generativity results from successful resolution of the conflict of intimacy versus isolation.
   b. Generativity usually occurs during early adulthood.
   c. Since intimacy is the focus of early adulthood, generativity usually occurs later, in middle adulthood.
   d. Generativity is caring for the next generation and helping to improve society.

3. Levinson's theory of development does NOT include (457-458)
   a. the life structure concept.
   b. a sequence of *eras* and *transitions.*
   c. dreams and mentors.
   d. many low-SES women.

4. Women's *dreams*, images of the self in the adult world, are likely (458)
   a. to be split between marriage and career.
   b. to be the same as men's.
   c. to be more individualistic than men's.
   d. to emphasize independence and career.

5. Mentors are (458)
   a. easier to find for women than for men.
   b. generally the same age as their students.
   c. are almost always friends.
   d. experienced in the world the person seeks to enter.

6. In Levinson's theory, "settling down" refers to the changes that take place during a person's 30s (459)
   a. concerning men's diminishing sex drive.
   b. concerning women's desire to marry and have children.
   c. concerning men's desire to establish a stable niche in society.
   d. concerning men's increased fidelity in their sexual relationships.

7. Vaillant's theory (459)
   a. contradicts Levinson's theory.
   b. fills in gaps between Erikson's stages.
   c. included the study of women, unlike Levinson's.
   d. is based on research conducted in the past 20 years.

8. The social clock (459-460)
   a. is the term coined for women's biological drive to reproduce.
   b. is a source of social comparison that may lead to low self-esteem.
   c. determines what is considered "fashionably late."
   d. governs how long people should stay at a party.

9. When it comes to selecting a mate (461)
   a. it is true that opposites attract.
   b. women prefer partners who are healthy and younger than themselves.
   c. men are primarily interested in physical attractiveness.
   d. women and men tend to have the same criteria for their selections.

10. Of the three components of love, which predicts a lasting relationship in the early stages of a relationship? (461-462)
   a. intimacy
   b. passion
   c. commitment
   d. shared attitudes and values

11. Couples who report having higher quality long-term relationships (462)
   a. have sex much more often than the average couple.
   b. somehow keep the mystery in their relationship.
   c. credit their success to being very different, because opposites attract.
   d. consistently communicate their commitment to each other.

12. Which is NOT a barrier to intimacy between male friends? (464)
   a. a feeling of competition among friends
   b. an unwillingness to disclose any weaknesses
   c. fear that friends will not reciprocate, if they tell about themselves
   d. fear that friends will gang up on anybody who is open

13. Who has the most other-sex friendships? (465)
   a. older women
   b. younger women
   c. married men
   d. single men

14. Which friendship is likely to last the longest? (464-465)
   a. one between women
   b. one between men
   c. one between brothers
   d. one between a man and a woman

15. Who is most likely to experience loneliness and under what conditions? (465)
   a. husbands (not wives) in a marriage
   b. women (not men) who are not in a romantic relationship
   c. single (not divorced) men or women
   d. mothers (not fathers) of small children

16. Which is a personal characteristic associated with persistent loneliness? (466)
   a. low intelligence
   b. insensitivity
   c. unattractiveness
   d. high intelligence

17. ______percent of American 18- to 24-year-olds reside with their parents. (466)
   a. 25
   b. 50
   c. 75
   d. 80

18. Of the young adults who leave home (466-467)
    a. some show signs of weakness by returning home.
    b. it is rare for them to return to their parents' home. Less than 25% do.
    c. many view the parental home as a safety net.
    d. those who leave because of family conflict are least likely to return.

19. _____percent of Americans marry at least once, and the age of first marriage has steadily ________________ over the last 50 years. (467)
    a. 75 percent, risen
    b. 85 percent, declined
    c. 90 percent, risen
    d. 95 percent, declined

20. Which of the following beliefs is true? (470)
    a. The best single predictor of marital satisfaction is the quality of the couple's sex life.
    b. If my spouse loves me, he or she should instinctively know what I want and need to be happy.
    c. No matter how I behave, my spouse should love me simply because he or she is my spouse.
    d. None of the above is true.

21. Cohabiting couples (476-477)
    a. may be testing their relationships for the rigors of marriage.
    b. may be seeking an alternative to marriage.
    c. have decrease in number since 1970.
    d. both a and b are true.

22. Which is NOT a factor that contributes to today's high divorce rates? (477-478)
    a. shifting societal conditions including widespread family poverty
    b. growing economic independence among women
    c. increasing numbers of men initiating divorce
    d. infidelity

23. Who is the most likely to experience conflict and poor adjustment to his or her parenting role following a divorce and remarriage? (479)
    a. stepmothers
    b. stepfathers
    c. biological mothers after remarriage
    d. biological fathers after remarriage

24. The most common never-married parents in their twenties are (479)
    a. Caucasian professional women who don't have time for a husband.
    b. African-American women, who give birth out of wedlock 60 percent of the time.
    c. Caucasian professional men left with children that their partners were unwilling or unable to care for.
    d. Hispanic women whose religion forbids the use of birth control.

25. Which is NOT a source of strain in dual-earner marriages as discussed in the text? (483)
    a. work overload, usually experienced by the wife
    b. feeling torn between excelling at their jobs and spending time with each other and their children
    c. little energy for socializing or entertaining
    d. conflict between spouses concerning whose job is better or more important

CHAPTER 15

PHYSICAL AND COGNITIVE DEVELOPMENT IN MIDDLE ADULTHOOD

BRIEF CHAPTER SUMMARY

During middle adulthood, the gradual physical changes that began in young adulthood continue. Age-related deterioration in visual abilities, hearing, and the condition of the skin become more apparent. Muscle mass declines, fat deposits increase, and bone density declines. Many physical changes are mediated by the effects of exercise and healthy lifestyles.

The climacteric, or decline in reproductive capacity, occurs over a 10-year period for women. Many doctors recommend hormone replacement therapy to counteract the negative effects of menopause, but it is controversial due to increased risk of cancer. Physical and psychological symptoms of menopause vary greatly between individuals and cultures. Sexual activity among married couples remains relatively stable in midlife.

Middle adulthood can bring many serious health concerns including cancer, cardiovascular disease, and osteoporosis. Hostility predicts health problems, while effective stress management (including both problem-centered and emotion-centered coping strategies), a healthy diet, regular exercise, social support, and regular medical screenings can reduce health risks considerably. An optimistic outlook is related to both physical and psychological well-being.

Cognitive changes during middle adulthood exemplify the lifespan view of development as multidimensional, multidirectional, and plastic. Crystallized intelligence (based on knowledge and experience) increases, while fluid intelligence (representing information-processing skills) declines. Slower processing speed makes effective attention more difficult, and reduction in the use of memory strategies contributes to declines in working memory. However, experience, practice, and metacognitive skills allow healthy adults to compensate for these deficits.

As in other periods of life, the relationship between vocational life and cognitive development during midlife is reciprocal: each builds on and contributes to the other. Life transitions often motivate a return to college, especially for women. Social support from family, friends, and institutional services can help returning students overcome the challenges they face and reap the benefits of expanding their horizons, as well as those of others.

LEARNING OBJECTIVES

After reading this chapter, you should be able to:

15.1. Describe physical changes of middle adulthood, paying special attention to vision, hearing, the skin, and muscle–fat makeup. (493-495)

15.2. Describe the reproductive changes that occur in women during middle adulthood, and discuss women's psychological reactions to menopause. (495-496)

15.3. Describe the reproductive changes in men during middle adulthood. (496)

15.4. Discuss sexuality in middle adulthood and its association with psychological well-being. (496-498)

15.5. Discuss cancer, cardiovascular disease, and osteoporosis, noting risk factors and interventions. (498-501)

15.6. Discuss the association of hostility and anger with heart disease and other health problems. (501-502)

15.7. Explain the benefits of stress management, exercise, and an optimistic outlook in dealing effectively with the physical challenges of midlife. (502--505)

15.8. Explain the double standard of aging. (505)

15.9. Describe changes in crystallized and fluid intelligence in middle adulthood, and discuss individual and group differences in intellectual development. (506-508)

15.10. Describe changes in information processing in midlife, paying special attention to speed of processing, attention, and memory. (508-510)

15.11. Discuss the development of practical problem solving, expertise, and creativity in middle adulthood. (511-512)

15.12. Describe the relationship between vocational life and cognitive development. (512-514)

15.13. Discuss challenges facing adults who return to college, ways of supporting returning students, and benefits of earning a degree in midlife. (514-516)

STUDY QUESTIONS

Physical Development

Physical Changes

1. List three declines in visual functioning resulting from thickening and yellowing of the lens, weakening of eye muscles, shrinking of the pupil, and reduced transparency of the vitreous. (493-494)
A. ______________________________
B. ______________________________
C. ______________________________

2. Age-related hearing difficulties are caused by both deterioration of ______________ structures that convert sound waves into ______________ and exposure to ______________. (494)

3. As we age, the ______________ becomes less firmly attached to the dermis, fibers in the ______________ thin, and fat in the ______________ diminishes. List resulting changes in the appearance of skin at the following ages: (494)
Thirties: ______________ Forties: ______________
Fifties: ______________

4. What environmental factor hastens wrinkling and spotting of the skin? (494)
____

5. True or False: Considerable weight gain and loss of muscle mass is inevitable in middle age. (494)

6. A gradual loss in bone mass begins in the _______________ and accelerates in the _______________, especially for (men/women). Weakened bones _________________ more easily and _____________ more slowly. (494-495)

7. In women, the climacteric occurs (rapidly/gradually) and concludes with ________________________, usually between ages _____ and _____. (495)

8. Benefits of hormone replacement therapy (HRT) include protection against ______ deterioration and ___________________ disease as well as reduction of certain discomforts such as ____________________ and ____________________. HRT may also improve ____________________ and other aspects of ______________________. However, it is associated with certain forms of ______________. (495)

9. List another way the benefits of HRT can be achieved. (495)
____
____

10. The way women react to menopause depends on how they interpret the event in relation to their __________________ and ______________ lives. (495)

11. Identify three contexts in which menopause can be viewed. Circle the letter of the one which is associated with the most negative attitudes and which is prevalent in industrialized nations. (497)
A. ___________________________________ B. ______________________________________
C. ___________________________________

12. Changes in women's cultural _________ also affect their tendency to experience negative physical and psychological symptoms at menopause. (497)

13. True or False: Testosterone production in men decreases with age. (496)

14. The inability to attain an erection when desired increases with age, affecting ______ percent of men by age 60. It is (likely/unlikely) to respond well to treatment. (496)

Health and Fitness

1. Men are more likely to be affected by _____________ illnesses, while women more often have ______________________ ones. (496)

2. Most studies of illness risk factors, prevention, and treatment have been carried out on _______. (496)

3. (Stability/Decline) in sexual activity is typical. The best predictor of sexual frequency is ________________________________. (496)

4. True or False: Both men and women take longer to feel aroused and reach orgasm during middle adulthood, but sex may become more satisfying. (497)

5. A higher male _______________ rate and the value women place on ____________________________________ in sexual relations make partners less available to them in midlife. (498)

6. List the two leading causes of death in midlife. (498)
A. ______________________________ B. ________________________________

7. Cancer occurs when the ______________________________ of a normal cell is disrupted, leading to uncontrolled growth and spread of abnormal cells. (498)

8. What three general factors contribute to cancer? (498)
A. ______________________________ B. ________________________________
C. ______________________________

9. (Women/Men) and (African Americans/Caucasian Americans) are more vulnerable to cancer. (499)

10. List ways to reduce the incidence of cancer and cancer deaths. (499)
____
____
____
____

11. Describe the relationship between SES and cancer. (499)
____
____
____

12. List the three signs of cardiovascular disease which often have no symptoms. (499)
A. ______________________________ B. ________________________________
C. ______________________________

13. List three symptoms of cardiovascular disease. (500)
A. ______________________________ B. ________________________________
C. ______________________________

14. List ways to reduce the risk of heart attack. (500)
____
____
____
____

15. Researchers have found that doctors are more likely to overlook (men's/women's) symptoms of cardiovascular disease. (500)

16. Osteoporosis affects ____ in four women and ________________ of both sexes over age 70, and may not be evident until _______________ occur. (500)

17. List risk factors for osteoporosis, in addition to age and sex. (500)
A. ______________________________ B. ________________________________
C. ______________________________ D. ________________________________
E. ______________________________ F. ________________________________
G. ______________________________

18. Increasing calcium and vitamin D intake and engaging in regular exercise in childhood, adolescence, and early adulthood reduce lifelong risk of osteoporosis by increasing ________________________________. (500-501)

19. Recent evidence pinpoints ________________ as the "toxic" ingredient of Type A personality, predicting ______________________________ as well as other health problems. (502)

20. True or False: Suppressing anger is a healthier way of dealing with negative feelings. (502)

Adapting to the Physical Challenges of Midlife

1. Identify several effective ways to manage stress. (502-503)

____
____
____
____

2. Identify and describe two general strategies people use to cope with stress. (503)
A.____
____
B.____
____

3. What is the best coping strategy? (503)

____

4. Give an example of a constructive approach to anger reduction. (503)

____

5. True or False: Community support services are less available for middle-age concerns than for those of young adulthood and the aging. (503)

6. Of adults who begin an exercise program, ______ percent discontinue within the first 6 months and less than ______ percent of those who continue exercise at levels that lead to health benefits. (504)

7. __________________________ helps people start exercise programs and stay with them. (504)

8. Identify characteristics of beginning exercisers that best fit group- versus home-based exercise programs. (504)
Group-based:____
____
Home-based:____
____

9. Hardy individuals view most experiences as ____________________, display a ____________________, involved approach to daily activities, and view change as ______________________________—a normal part of life and a chance for personal growth. (504)

10. The positive appraisals of stressful situations resulting from characteristics of hardiness predict ____________________ behaviors, tendency to seek ______________________, fewer ____________________, and ___________________–centered coping in controllable situations. characteristics, a gap that widens with age. (504)

11. Middle-aged (men/women) are rated as less attractive and as having more negative characteristics, a gap that widens with age. (Men/Women) judge an aging woman more harshly. (505)

12. (Men/Women) judge an aging woman more harshly. (505)

13. What do researchers believe is at the heart of the double standard of aging? (505)

____

Cognitive Development

Changes in Mental Abilities

1. What core principles of the lifespan perspective do cognitive changes during middle adulthood exemplify? (506)

A. ________________________________ B. ________________________________

C. ________________________________

2. Crystallized intelligence, representing accumulated ____________________ and good __________________, is made up of capacities that are acquired because they are ____. (507)

3. Fluid intelligence, representing ____________ of processing and ___________ capacity, is believed to be influenced more by ______________________________ ______________________________________ and less by ______________. (507)

4. ______________________ intelligence increases steadily with age into late adulthood, while ______________________ intelligence begins to decline in the late twenties and early thirties. (507)

5. List five mental abilities which tap both crystallized and fluid intelligence and one that taps only fluid intelligence. (507-508)

A. ________________________________ B. ________________________________

C. ________________________________ D. ________________________________

E. ________________________________ Fluid: ____________________________

6. List three reasons why abilities that tap crystallized and fluid intelligence tend to remain stable during middle adulthood despite information-processing declines. (508)

A. ____

B. ____

C. ____

7. True or False: Continued use of intellectual skills does not seem to contribute to their maintenance. (508)

8. Briefly describe the gender differences in decline of mental abilities. (508)

____

____

9. What explains the better performance on verbal memory, inductive reasoning and spatial orientations by the baby boom generation compared with the previous generation? (508)

____

10. What explains the decline in numeric ability? (508)

____

Information Processing

1. Two explanations exist for the decline in cognitive functioning. According to the neural network view, ____________ occur in neural networks as neurons ______, requiring bypasses. The information-loss view suggests that information ______________________ more quickly in older adults, requiring more time for interpretation. (509)

2. True or False: Even familiar tasks are performed with less speed in older individuals. (509)

3. The ability to attend to two activities at once and ignore irrelevant information (increases/decreases) with age. (509)

4. What attention-related change in visual information processing occurs? (510)

____

5. As in the case of processing speed, _________________ and _________________ reduce the decline of attentional abilities. (510)

6. What explains the decline in working memory from the twenties into the sixties? (510)

____

7. Many middle-aged adults are not motivated to use memory strategies because they are not _________________. Their performance is improved when they are instructed to use strategies or when information is presented at a __________________________. (510)

8. Factual knowledge, procedural knowledge, and occupational knowledge ____________________________ in middle adulthood, and _________________ knowledge can be used to compensate for difficulty in recall. (510)

9. What can help us understand why practical problem solving takes a leap forward in middle adulthood? (511)

____

10. The development of expertise is at its peak in _________________, leading to highly effective problem solving organized around __________________ principles and _________________ judgments. (511)

11. True or False: Expertise is only evident in careers requiring advanced education. (511)

12. Everyday problem solving peaks between ages _____ and _____ and may be sustained longer. (511)

13. With advancing age, the focus of creativity may shift from the generation of unusual products to the integration of _____________________ and ___________________ into unique solutions, as well as from ______________ concerns to more ______________ goals. (512)

14. True or False: When given a challenging, real-world problem, related to an area of expertise, the middle-aged adult is likely to perform better than younger adults not only in efficiency, but in quality and originality of thinking. (512)

Vocational Life and Cognitive Development

1. The relation between vocational life and cognition is __________________. Such findings in Japan and Poland help explain the relationship between social class and ________________ thinking, as well as peoples' work and their choice of ______________________ pursuits. (514)

2. True or False: Cognitive flexibility is less responsive to vocational pursuits in middle age than it is in young adulthood. (514)

Adult Learners: Becoming a College Student in Midlife

1. Life _______________ often trigger a return to formal education. (514)

2. The majority of adult learners are _____________. (514)

3. Describe the typical situation of women returning to college who report the greatest psychological stress. (514-515)

____

____

4. List ways to support and facilitate adult reentry to college. (515)

____

____

____

____

5. List benefits of adult reentry to college. (515-516)

____

____

ASK YOURSELF . . .

When Stan was 42, his optician told him that he needed bifocals, and over the next 10 years, Stan required an adjustment to his corrective lenses almost every year. What physical changes account for Stan's repeated need for new eye wear? (see text pages 493-494)

____

____

____

Nancy noticed that between ages 40 and 50, she gained 20 pounds, and her arm and leg muscles seemed weaker. She had trouble opening tightly closed jars, and her legs ached after she climbed a flight of stairs. Nancy thought to herself, "Exchanging muscle for fat must be an inevitable part of aging." Is Nancy correct? Why or why not? (see text page 494)

Cite cultural and gender-role influences on the experience of menopause. (see text pages 495-497)

When Cara complained of chest pains to Dr. Furrow, he decided to "wait and see" before conducting additional tests. In contrast, Cara's husband Bill received a battery of tests aimed at detecting cardiovascular disease, even though he did not complain of symptoms. What might account for Dr. Furrow's different approach to Cara than Bill? (see text pages 499-500)

Because his assistant misplaced some important files, Tom lost a client to a competitor. Tom felt his anger building to the breaking point. Explain why Tom's response is unhealthy, and suggest effective ways for dealing with it. (see text pages 502-503)

According to the lifespan perspective, development is multidimensional—affected by biological, psychological, and social forces. Explain how this is so for health at midlife, citing important influences. (see text pages 496-504)

In what aspects of cognition did Devin decline, and in what aspects did he gain? How do changes in Devin's thinking reflect assumptions of the lifespan perspective? (see text page 506)

Asked about hiring older adults as waiters, one restaurant manager replied, "I cannot hire enough older workers . . . they are my best employees . . . I do not even know what you mean by slowness. They are the fastest ones on the floor" (Perlmutter, Kaplan, Nyquist, 1990, p. 189). Why does this manager find older employees desirable, despite the age-related decline in speed of processing? (see text page 509)

Consider the famous saying, " You can't teach an old dog new tricks." Evaluate its accuracy in terms of evidence on the impact of vocational and educational experiences on cognitive development in midlife. (see text pages 512, 514)

____

____

____

Why do most high-level government and corporate positions go to middle-aged and older adults rather than to young adults? What cognitive capacities enable mature adults to perform well in these jobs? (see text pages 511-512)

____

____

____

Marcella completed only one year of college in her twenties. Now, at age 42, she returned to earn a degree. Plan a set of experiences for the month before Marcella enrolls and her first semester that will increase her chances of success. (see text page 514-515)

____

____

____

SUGGESTED READINGS

Buunk, B. P., & Gibbons, F. X. (Eds.). (1997). *Health, coping, and well-being: Perspectives from social comparison theory*. Mahwah, NJ: Erlbaum. Applies social comparison theory to a wide range of health-related topics, including health protective behaviors, risk behaviors, coping with serious disease, perceptions of risk, seeking medical care, and depression.

Kotre, J. (1995). *White gloves: How we create ourselves through memory.* New York: Free Press. Explores autobiographical memories as a means of archiving our past experiences. Memory distortion is discussed as an aid in creating meaning about the self. Includes historic and recent research on autobiographical memory. The author records the stories of people's lives, and many of these stories have contributed to the production of the *Seasons of Life* video series.

McCabe, P. M., Schneiderman, N., Field, T. M., & Wellens, A. R. (Eds.). (1999). *Stress, coping, and cardiovascular disease*. Mahwah, NJ: Erlbaum. Discusses physiological, psychosocial, developmental, and mental health factors in the relationship between stress and heart disease.

Schacter, D. L. (1996). *Searching for memory: The brain, the mind, and the past.* New York: Basic Books. Synthesizes research on memory. Discusses the subjective nature of memory, how memories are encoded and retrieved, memory loss, the controversy concerning repressed memories, and what happens to memory as we age.

PUZZLE 15.1 TERM REVIEW

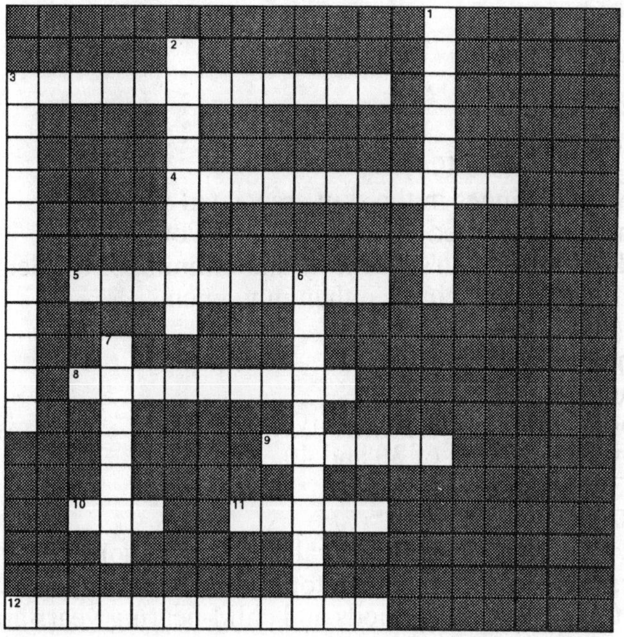

Across

3 Intellectual skills that depend on accumulating knowledge, experience, good judgment, and mastery of social conventions

4 ____-loss view: attributes age-related slowing of cognitive processing to greater loss of information as it moves through the system

5 Condition of aging in which the lens of the eye loses its capacity to accommodate to nearby objects

8 End of menstruation and reproductive capacity in women

9 _____ _ behavior pattern: extreme competitiveness, impatience, hostility, and a sense of time pressure (2 words)

10 Therapy for women that involves daily doses of estrogen (abbr.)

11 Intellectual skills that depend largely on basic information-processing skills--speed of analysis, working memory capacity, and detection of visual pattern relationships

12 Severe version of age-related bone loss in which porous bones fracture easily and, when severe, lead to a slumped-over posture and shuffling gate

Down

1 _____ problem solving: sizing up real-world situations and analyzing how best to achieve goals that have a high degree of uncertainty

2 Three personality characteristics--control commitment, and challenge--that help people cope adaptively and reduce the the impact of stress on health

3 Midlife transition in which fertility declines

6 Age-related hearing impairments involving a sharp loss at high frequencies around age 50 that gradually extend to all frequencies

7 Neural _____ view: attributes age-related slowing of cognitive functioning to breaks in neural networks as neurons die

SELF-TEST

1. Vision problems associated with aging include all of the following EXCEPT (494-495)
   a. presbyopia.
   b. yellowing lenses.
   c. shrinking lenses.
   d. light sensitivity.

2. Wrinkles and "age spots" (494)
   a. are uncommon changes in the skin associated with aging.
   b. can be somewhat reduced by tanning while young.
   c. are related to loss of skin's elasticity and thinning of the dermis.
   d. appear at a younger age in men than in women.

3. Women's climacteric (495)
   a. occurs gradually over one year.
   b. does not bring reproductive capacity to an end.
   c. begins between the ages of 30 and 40.
   d. concludes with menopause.

4. Menopause evokes which psychological reaction in women? (495)
   a. unhappiness due to physical symptoms and loss of sex appeal
   b. relief that their menstrual periods and child-bearing years are over
   c. less favorable feelings for Black and Hispanic women than for Caucasians
   d. both a and b

5. Reproductive capacity for men (496)
   a. stays the same throughout their lives.
   b. ends with the inability to attain an erection common between 60 and 90.
   c. ends with male menopause and an end to sperm production.
   d. diminishes after age 40 but remains possible throughout their lives.

6. Women are more likely than men (496-501)
   a. to suffer from fatal illnesses during middle adulthood.
   b. to suffer from osteoporosis because of lower bone mineral reserves.
   c. to consider their health to be either "poor" or "excellent."
   d. to suffer from cardiovascular disease in middle adulthood.

7. Sexual activity during middle adulthood (496)
   a. typically increases, but only slightly.
   b. typically declines, but only slightly.
   c. is not linked to emotional intimacy.
   d. is more likely to decline for men than it is for women.

8. Cancer (498)
   a. is by far the leading cause of death for middle-aged women.
   b. is responsible for about 25 percent of all deaths in midlife each year in the United States.
   c. patients are cured (free of the disease for 5 or more years) 40 percent of the time.
   d. is both a and c.

9. Hostility (501-502)
   a. is a consistent predictor of heart disease and other health problems.
   b. does not negatively affect one's health as long as it is not kept inside.
   c. is only dangerous when it leads to unhealthy behaviors such as smoking, alcohol consumption, and overeating.
   d. is both b and c.

10. Which of the following is NOT an effective health intervention? (501-504)
   a. stress management
   b. exercise
   c. an optimistic outlook when dealing with midlife changes
   d. emotion-centered and avoidant coping strategies

11. Which is NOT a barrier to beginning an exercise regimen during midlife? (504)
   a. being part of an exercise group
   b. lack of time and energy
   c. inconvenience
   d. work conflicts

12. Which is NOT a personal quality associated with hardy individuals? (504)
   a. control
   b. strength
   c. challenge
   d. commitment

13. The broader culture views middle-aged (505)
   a. women as assertive, confident, and versatile.
   b. men as unattractive, weak, and feeble minded.
   c. women as unattractive, incompetent, and passive.
   d. men and women, alike, as feeble minded, "washed-up," and unattractive.

14. Crystallized intelligence (507)
   a. is influenced by conditions of the brain and learning unique to the individual.
   b. does not depend on accumulated knowledge and experience.
   c. depends heavily on basic information-processing skills.
   d. represents abilities that are valued by an individual's culture.

15. Fluid intelligence (507)
   a. depends on accumulated knowledge and experience.
   b. improves in the late twenties or early thirties.
   c. is tested by intelligence test items such as vocabulary, general information, verbal analogy, and logical reasoning.
   d. is based heavily on the speed with which we can analyze information, the capacity of working memory, and the ability to detect relationships among stimuli.

16. Research concerning the stability of cognitive functioning reveals (508)
   a. that middle-age declines in mental abilities are inevitable and severe.
   b. that the decrease in basic processing speed is modest until late in life.
   c. that unfamiliar tasks are easier for older adults to accomplish than repetitious ones.
   d. both b and c.

17. According to the text, which factor does NOT affect the degree to which intellectual skills are retained? (508)
    a. watching less than 4 hours of television per day
    b. being physically fit and free from disease
    c. being in a lasting marriage
    d. traveling

18. Reaction time (509)
    a. slows at a faster rate for women than men.
    b. slows several seconds during middle adulthood.
    c. generally begins getting slower when individuals are in their forties.
    d. does not change across the lifespan.

19. During middle and late adulthood (510)
    a. working memory increases from the twenties into the sixties.
    b. it becomes more difficult to ignore irrelevant stimuli.
    c. factual and procedural knowledge remain unchanged or increase.
    d. both b and c are true.

20. The development of expertise (511)
    a. reaches its height during midlife.
    b. has no real effect on practical problem solving abilities.
    c. emerges only in individuals with advanced educations.
    d. both a and c are true.

21. High-level, integrative thinking in any context (511-512)
    a. peaks in the early college years.
    b. is dependent on educational level.
    c. peaks between age 40 and 59.
    d. is related to IQ.

22. Vocational life (512, 514)
    a. is not related to cognitive development.
    b. is affected solely by personality characteristics
    c. is not an influence on the cognitive growth of a 50-year-old.
    d. choices are influenced by cognitive flexibility.

23. Adult students, over the age of 25, represent nearly _____ of all college students (514)
    a. 50 percent
    b. 10 percent
    c. 25 percent
    d. one-third

24. Students over the age of 35 (514)
    a. are mostly women.
    b. are mostly men.
    b. report less stress than traditional 18 to 24 year-old students.
    d. are very active participants in class discussions.

25. Returning to college during middle adulthood (515-516)
    a. usually ends in dissatisfaction with coursework and dropping out, even when support systems are in place.
    b. increases unfavorable stereotypes held by traditional students.
    c. is especially difficult because of the need to form new personal relationships and share opinions and experiences.
    d. can reshape the course of development for these adults

CHAPTER 16

EMOTIONAL AND SOCIAL DEVELOPMENT IN MIDDLE ADULTHOOD

BRIEF CHAPTER SUMMARY

Concerns about making meaningful contributions to family and society increase greatly during middle adulthood, consistent with Erikson's psychological conflict of generativity versus stagnation. According to Levinson, middle-aged adults reassess their relation to themselves and the world and make adjustments in areas that are not satisfactory. Vaillant added that middle-aged adults become guardians of the traditions and values of their culture. Few people experience midlife as a crisis, although most must adapt to important events that lead to new understandings and goals.

Self-concept changes during middle age, reflecting fewer and more concrete possible selves, a growing awareness of a finite lifespan, longer life experience, and generative concerns. Adults become more introspective and self-acceptance, autonomy, environmental mastery, and coping strategies improve. Both men and women become more androgynous, as well. Despite such changes in the organization of personality, little change takes place in basic, underlying personality traits during midlife and beyond.

Launching children brings important changes to which most parents adjust well. Many adults focus on improving their marriages during these years. When divorce occurs, it is more difficult for middle-aged adults than for younger adults. As long as family relationships are positive, becoming grandparents facilitates development.

Many middle-aged adults become squeezed between the needs of aging parents and financially dependent children. While middle-aged adults often become more appreciative of the contribution of their own parents, caring for ill parents can be very stressful. Sibling ties, which often continue to strengthen, can help lessen the burden as can other social supports. Friendships become fewer and more selective in midlife. Contrary to popular belief, research reveals strong inter-generational ties and feelings of support in U.S. society.

Middle-aged adults seek to increase the personal meaning and self-direction of their work. Overall job satisfaction improves during midlife, but burnout has become a greater problem in recent years, especially in careers involving work with people. Vocational development is less available to older workers and many women and ethnic minorities leave the corporate world to escape the glass ceiling, which limits their advancement. Still, radical career changes are rare in middle adulthood. Unemployment is especially difficult for middle-aged individuals, and retirement is an important change that is often stressful, making effective planning important for positive adjustment.

LEARNING OBJECTIVES

After reading this chapter, you should be able to:

16.1. Describe Erikson's stage of generativity versus stagnation and related research findings. (522-524)

16.2. Describe Levinson's and Vaillant's views of psychosocial development in middle adulthood, and discuss similarities and differences in midlife changes for men and women. (525-527)

16.3. Discuss the extent to which midlife crisis captures most people's experience of middle adulthood. (527-528)

16.4. Characterize middle adulthood using a life events approach and a stage approach. (528)

16.5. Describe stability and changes in self-concept and personality in middle adulthood, including coping styles and gender identity. (528-533)

16.6. Describe the middle adulthood phase of the family life cycle, and discuss midlife relationships with a marriage partner, adult children, grandchildren, and aging parents. (534-541)

16.7. Describe midlife sibling relationships and friendships, and discuss relationships across generations in the United States. (541-543)

16.8. Discuss job satisfaction and vocational development in middle adulthood, paying special attention to the experiences of women and ethnic minorities. (543-546)

16.9. Discuss career change and unemployment in middle adulthood. (546-547)

16.10. Discuss the importance of planning for retirement, noting various issues that middle-aged adults should address. (547-548)

STUDY QUESTIONS

Erikson's Theory: Generativity versus Stagnation

1. What does generativity involve in middle age? (522-523)

____

2. List four ways generativity can be expressed in addition to the bearing and raising of children. (523)

A. ____________________ B. ____________________
C. ____________________ D. ____________________

3. Erikson believed that the belief that life is basically ____________________ ____________________ is the underlying motivation for generative action. (523)

4. What is strongly linked with dedication to political causes in midlife? (524)

____

5. People with a sense of stagnation are unable to contribute to society's welfare because they place ____________________ above challenge and sacrifice. (523)

6. True or False: Research indicates that generativity does increase during middle adulthood. (524)

7. List several characteristics of highly generative people. (524)
____
____
____

Other Theories of Psychosocial Development in Midlife

1. Around age ______, people evaluate their success at meeting early adulthood goals and regard the remaining years as increasingly ________________. (525)

2. At midlife, adults strive to create a ____________________ balance more in tune with their time of life. (525)

3. As middle-aged adults confront mortality, they become more aware of ways people act ____________________ and counter this force with a desire to be __________________ through endeavors that advance human welfare. (526)

4. During middle age, people also reconcile _______________________________ and ___________________________ parts of the self. (526)

5. Midlife requires a middle ground between _____________________________ and ____________________. Men and career-oriented women generally (increase/reduce) their concern with achievement, while those previously devoted to child rearing or unrewarding careers do the opposite. (526)

6. Name three contexts that forestall the pursuit of a satisfying life structure in favor of survival. (526)
A. ________________________________ B. __________________________________
C. ________________________________

7. For what two groups are opportunities for advancement, which permit the realization of the early adulthood dream and ease the transition to middle adulthood, less available? (526)
A. ____________________ B. ____

8. True or False: Nearly twice as many women as men are illiterate. (527)

9. Older people in societies around the world are guardians of ______________ ____________________________________, holding rapid change in check. (527)

10. Research suggests that there is (little/great) diversity in response to midlife. Changes for men are more likely to occur in the ________________ and those for women in the _________________________. (527)

11. True or False: Midlifers who change careers typically view the change as a radical shift. (527-528)

12. What three factors typically limit the early adulthood satisfaction of the few individuals who experience crisis at midlife? (528)
A. ________________________________ B. __________________________________
C. ________________________________

13. Psychosocial changes coincide with both ________________________ events and ________________________ age. (528)

14. Like other periods, midlife is a period of both __________________ and ________________________ change. (528)

Stability and Change in Self-Concept and Personality

1. As we age, we may evaluate ourselves less according to ________________ comparisons and more according to how well we measure up in relation to what we ________________________________. (528)

2. How do possible selves change with age? (529)
____
____
____

3. In what important way are possible selves different from current self-concept? How does this impact self-esteem during middle age? (529)
A. ____
B. ____

4. One of the most consistent personality changes in midlife is a rise in ________________________________, as people contemplate the second half of life. (529)

5. What other traits increase from early to middle adulthood in well-educated individuals? (529)
A. ________________________ B. ________________________
C. ________________________

6. List 5 factors that influence psychological well-being at midlife. (530-531)
A. ________________________ B. ________________________
C. ________________________ D. ________________________
E. ________________________

7. True or False: Notions of happiness are relatively the same in all cultures. (529)

8. Describe differing coping mechanisms of middle-aged and younger men in Vaillant's sample. (529-531)
Middle-aged:____
____
____
Younger:____
____

9. In addition to greater self-acceptance and confidence, what may contribute to more sophisticated, flexible coping during middle age? (531)
____

10. Gender identity becomes more ________________________ in midlife. (531-532)

11. The above change is the product of what two forces? (532)
A. ______________________________ B. ______________________________

12. List the "big five" personality traits. (533)
A. ______________________________ B. ______________________________
C. ______________________________ D. ______________________________
E. ______________________________

13. What three personality traits tend to decline from the teenage years to the end of the twenties, and which two increase? (533)
Decline: ______________________________
Increase: ______________________________

14. True or False: While people's strategies and goals change, underlying personality traits are relatively stable. (533)

Relationships at Midlife

1. Compared to other age groups, middle-age households are financially ______________________________. (534)

2. ______________________________ is a strong contributor to the ______________________________ of poverty. The gender gap in poverty in the United States is (lower/the same/greater) than in other Western industrialized nations. (535)

3. List four common reasons given by women for divorce during middle age. (535)
A. ______________________________ B. ______________________________
C. ______________________________ D. ______________________________

4. Middle-aged women who successfully weather divorce tend to become more ______________________________, ______________________________, and ______________________________. (535)

5. A (minority/majority) of middle-aged parents have difficulty adjusting to the launching of their children. (535)

6. A strong __________ orientation predicts gains in life satisfaction when children leave the home. (535)

7. Adolescents and young adults who are ______________________________ in development can lead to parental strain. (535)

8. Departure of children is a relatively minor event when ______________________________ are sustained. (536)

9. Providing emotional and financial support to children during midlife is related to ______________________________. (536)

10. Once children strike out on their own, members of the middle generation, especially ______________________________, usually take on the ______________________________ role, gathering family members for celebrations and making sure they stay in touch. (536)

11. List some ways middle-aged parents can promote positive ties with their adult children. (536)

____
____
____
____

12. The average age of grandparenthood for women is ______________ and for men is ____________. (537)

13. List four commonly mentioned gratifications of becoming a grandparent. (537)
A. ____
B. ____
C. ____
D. ____

14. Relationships are typically closest between grandparents and grandchildren of the same sex—especially ___________________________ and ____________________ ________________________________. (537)

15. List subcultures in which grandparents—especially the maternal grandmother—become active in child rearing. (537-538)

____
____

16. Grandparents have increasingly stepped in as ______________ caregivers in the face of serious family problems. (538)

17. Briefly describe some of the challenges grandparents raising their grandchildren encounter. (538-539)

____
____
____

18. (More/Fewer) aging adults live with younger generations now than in the past. Frequency of contact is (high/low) and proximity (increases/decreases) with age. (538)

19. Many adults become (more/less) appreciative of their parents strengths and generosity. Parent-to-child help-giving ______________ while child-to-parent help-giving ______________. (538)

20. When an aging parent's spouse cannot provide care, ___________________ are the next most likely relatives to do so. (540)

21. As adults move from early to later middle age, sex differences in parental caregiving (increases/declines). (540)

22. The need for parental care typically occurs (gradually/suddenly), adult children usually feel a sense of grief over the loss of _____________________________, and caregivers often feel they no longer have __________________________________ __________________. (540-541)

23. List four ways to relieve the stress of caring for an aging parent. (541)
A. ______________________________
B. ______________________________
C. ______________________________
D. ______________________________

24. Sibling relationships that were previously positive often become (closer/more distant) during middle adulthood. (542)

25. List three factors that influence the frequency of adult sibling contact in Western nations. (542)
A. ______________________ B. ______________________
C. ______________________

26. In contrast to Western sibling relationships in which ties between __________ are often closest, Asian Pacific Islander ______________ attachments are strong and central to family social life. (542)

27. When couples gather with friends, why do men often congregate in one area and women in another? (542)
____

28. Number of friends (increases/declines) with age. (542)

29. Viewing a spouse as a ______________________ contributes greatly to marital happiness. (542)

30. (Strong/Moderate/Weak) connections between age groups within the family and community exist in the United States and most people (are/are not) guided by self-interest in how they think government resources should be distributed. (543)

Vocational Life

1. More so than at other ages, middle-aged workers strive to increase the ______________ and ________-direction of their vocational lives. (544)

2. High ______________ and low ______________ job satisfaction is linked to the emotional exhaustion and loss of personal control aspects of burnout, while the reverse is associated with feelings of reduced accomplishment. (544)

3. Less ______________________ are available to older workers, and older employees are less likely to participate in such activities when they are offered. (545)

4. With age, __________ needs decline in favor of ______________ needs, and older workers receive more ______________ tasks which do not facilitate updating. (545)

5. True or False: The low numbers of women and minorities in executive positions can be attributed to poor management skills. (545-546)

6. Women (do/do not) demonstrate qualities linked to leadership and advancement. (546)

7. When an extreme career shift occurs, it usually signals a personal ______________. (547)

8. As companies downsize and jobs are eliminated, the majority of those affected are ______________________________, who experience greater ______________________________________ as a result of job loss. (547)

9. ___________-centered coping skills can help with the adjustment to job loss, as can social support focusing on the person's ______________ and communication with others who share similar interests. (547)

10. Increasing numbers of people spend up to ____________ of their lives in retirement. (547)

11. _______________ leads to better retirement adjustment and satisfaction. (547)

12. List several components of an effective retirement plan. (548)

____

____

ASK YOURSELF . . .

Explain how Elena's decision to change careers is a joint product of inner needs and external pressures. (see text pages 525-528)

____

____

____

After years of experiencing little personal growth at work, 42-year-old Mel started to look for a new job. When he received an attractive offer in another city, he felt torn between leaving friendships built over many years and a long-awaited career opportunity. After several weeks of soul searching, he took the new job. Was Mel's dilemma a midlife crisis? Why or why not? (see text pages 525-528)

____

____

____

Around age 40, Luellen no longer thought about becoming a performing pianist. Instead, she decided to concentrate on accompanying other musicians in her community and expanding her studio of young pupils. How do Luellen's plans reflect changes in possible selves at midlife? (see text pages 528-529)

____

____

____

On his fifty-second birthday, Tom was asked how he felt now that another year had gone by. He replied, "I feel calmer and more content than at any time in my life." What personality changes might have contributed to Tom's response? (see text page 529)

____

____

____

Jeff, age 46, suggested to his wife, Julia, that they set aside a special time each year to discuss their relationship—both positive aspects and ways to improve. Julia reacted to Jeff's suggestion with surprise, since he had never before seemed interested in working on the quality of their marriage. What developments at midlife probably fostered this new concern? (see text pages 531-532)

____

____

____

Freda divorced her husband after 25 years of marriage. Compared to her daughter Salena, who got divorced at age 30, Freda had a harder time adjusting. Explain why adapting to divorce is more difficult for middle-aged than young adults and is particularly stressful for women. (see text pages 534-535)

____

____

____

Raylene and her brother Walter live in the same city as their aging mother, Elsie. When Elsie could no longer live independently, Raylene took primary responsibility for her care. What factors probably contributed to Raylene's involvement in caregiving and Walter's lesser role? (see text pages 540-541)

____

____

____

As a young adult, Daniel maintained close friendship ties with six college classmates. At age 45, he continued to see only two of them. What explains Daniel's reduced circle of friends in midlife? (see text page 542)

____

____

____

Ira recalls complaining about his job far more in early than middle adulthood. At present, he finds his work interesting and gratifying. Describe factors that probably caused Ira's job satisfaction to increase with age. (see text pages 544-545)

____

____

____

Trevor assigned the older members of his work team routine tasks because he was certain they could no longer handle complex assignments. Cite evidence presented in this and the previous chapter that shows Trevor is wrong. (see text pages 512, 514, 545)

____

____

____

An executive asks you what his large corporation ought to do to promote advancement of women and ethnic minorities to upper management positions. What would you recommend? (see text pages 545-546)

____

____

____

SUGGESTED READINGS

Booth, A., & Crouter, A. C. (Eds.). (1998). *Men in families: When do they get involved? What difference does it make?* Mahwah, NJ: Erlbaum. Discusses research on men's involvement in family relationships and their various familial roles as husband, father, and provider.

Cicirelli, V. G. (1995). *Sibling relationships across the life span.* New York: Plenum. Examines sibling relationships from childhood through adulthood, presenting cross-cultural evidence for both universals and variations in sibling ties. Also discusses a variety of special aspects of sibling relationships, including the impact of chronic illness and disability, conflict, violence and abuse, incest and sexual abuse, and death.

Ganong, L. H., & Coleman, M. (1999). *Changing families, changing responsibilities: Family obligations following divorce and remarriage.* Mahwah, NJ: Erlbaum. Examines shifts in the rights and duties of individual family members following changes in family structure caused by divorce and remarriage.

Kroger, J. (1999). *Identity development: Adolescence through adulthood.* Thousand Oaks, CA: Sage. Examines the development of identity from adolescence through late adulthood.

Turner, B. F., & Troll, L. E. (Eds.). (1994). *Women growing older: Psychological perspectives.* Thousand Oaks, CA: Sage. Presents a lifespan perspective on women's aging. Chapters include information on goal orientations across the life cycle, the social clock, gender stereotypes, creativity, attachment, social relations, and the relevance of psychodynamic theories for understanding gender among older women.

PUZZLE 16.1 TERM REVIEW

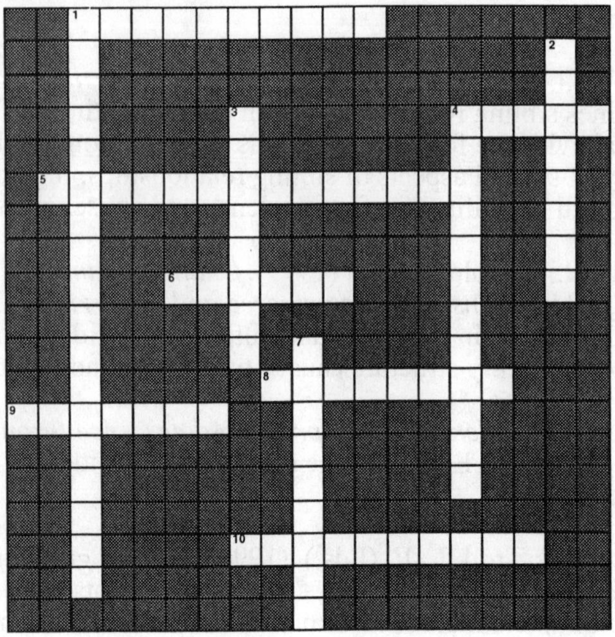

Across

1 Negative outcome of Erikson's conflict of midlife

4 _____ ceiling: invisible barrier to career advancement faced by women and minorities

5 _____ of poverty is a trend in which female heads of household have become the majority of the adult poverty populations, regardless of age and ethnicity.

6 Midlife _____: inner turmoil, self-doubt, and major personality restructuring at the transition to middle adulthood; characterizes few adults

8 Neuroticism, extraversion, openness to experience, agreeableness, and conscientiousness, called the "____ ____" personality traits (2 words)

9 Condition resulting from long-term job stress leading to emotional exhaustion, a sense of loss of personal control, and feelings of reduced accomplishment

10 Parental _____ theory: biological view claiming that traditional gender roles are maintained during active parenting years to ensure survival of children and become flexible after children reach adulthood

Down

1 Families in which children live with their grandparents, apart from their parents (2 words)

2 _____ selves: future-oriented representations of what one hopes to become and is afraid of becoming

3 _____ generation: middle-aged adults with ill or frail parents typically facing competing demands of their own children

4 Positive resolution of Erikson's conflict of midlife; integration of personal goals with the welfare of society

7 Role assumed by middle generation members (usually mothers) who take responsibility for family gatherings and connection

SELF-TEST

1. Erikson's stage of generativity versus stagnation does NOT involve (522-523)
   a. reaching out to others in order to guide the next generation, if generativity is attained.
   b. a combination of inner desires and cultural demands.
   c. self-absorption, self-centered, and self-indulgent behavior that is the negative outcome, stagnation.
   d. the formation of a coherent set of values and life plans.

2. Which is NOT a personality trait considered to be a component of generativity? (523-524)
   a. assertiveness
   b. competitiveness
   c. responsibility
   d. nurturance

3. Which of the following is NOT true of highly generative people? (523-524)
   a. They tend to be psychologically fulfilled and healthy.
   b. They are open to differing viewpoints and possess leadership qualities.
   c. They care greatly about the welfare of their children, partner, and society.
   d. They are self-centered and self-indulgent.

4. According to Levinson (525-526)
   a. virtually all people experience considerable confusion as they start to modify the components of their lives after midlife reassessment.
   b. midlife transition is a depressing time for all adults.
   c. drastic changes such as divorce and remarriage are unlikely to result from midlife reassessment.
   d. middle-aged adults realize that they have lived only half of their lives, so they have a great deal of time to accomplish their goals.

5. Women (525-526)
   a. are less likely than men to perceive themselves as younger than their chronological age.
   b. find it easier than men to accept being older.
   c. are less open to "masculine" characteristics such as autonomy, dominance, and assertiveness during middle age.
   d. are likely to try numerous remedies to maintain their youth, because of their desperate fear of becoming unattractive and unlovable.

6. According to Vaillant (527)
   a. well adjusted men and women enter a calmer, quieter time of life when they reach their fifties.
   b. adults are not ready to "pass the torch" until they are very late in life.
   c. rapid changes in cultural values initiated by young adults makes older adults lose touch with society and subsequently withdraw.
   d. adults focus more on personal goals as they move toward the end of middle age.

7. Midlife crisis is (527-528)
   a. a period of substantial inner turmoil for most men and women, according to Levinson.
   b. a period of slow and steady change, not crisis, according to Valliant.
   c. likely to occur earlier in women (early forties) than in men (late forties).
   d. both a and b.

8. Middle adulthood is (528)
   a. simply an adaptation to life events, not a stage.
   b. a period of crisis and major restructuring for all adults.
   c. characterized by discontinuity.
   d. a stage because the large majority of middle-aged people report troubling moments that prompt new understandings and goals.

9. Possible selves (528)
   a. are representations of what one is afraid of becoming.
   b. are representations of what one hopes to become.
   c. are more numerous in younger adults who rely less on social comparison than do middle adults.
   d. are both a and b.

10. Which of these traits does NOT increase from early to middle adulthood? (529)
    a. ambition
    b. environmental mastery
    c. autonomy
    d. self-acceptance

11. Which of the following factors does NOT promote psychological well-being at midlife? (530-531)
    a. sense of control over health, family, and work
    b. gratifying social ties
    c. exercise
    d. separation or divorce

12. Effective coping strategies are increasingly used in middle adulthood. For example, middle-aged men are more likely to (530-531)
    a. not care about problems.
    b. "ride" problems out.
    c. look for the "silver lining."
    d. avoid using humor to express ideas and feelings.

13. Gender identity (531-532)
    a. becomes more androgynous in midlife.
    b. becomes less androgynous in midlife.
    c. becomes more masculine for men during middle adulthood.
    d. becomes more feminine for women during middle adulthood.

14. Parental imperative theory suggests that __________ to help ensure the survival of the children. (532)
    a. men emphasize nurturance of wife and children
    b. women become more goal-oriented
    c. parents express the "other-gender" side of their personalities during active parenting years
    d. traditional gender roles are maintained during active parenting years

15. Which of the following “big five” personality traits change the most between the ages of 30 and 60? (533)
    a. extroversion
    b. openness to experience
    c. conscientiousness
    d. none of the “big five” change much at all

16. The middle adulthood phase of the family life cycle is (534)
    a. referred to as “launching children and moving on”.
    b. the last and shortest phase in the family life cycle.
    c. a period marked by few exits and entries of family members.
    d. financially a comparatively tough time for families.

17. In the United States, divorces during midlife (534)
    a. characterize 40 percent of marriages lasting 20 years or longer.
    b. are well accepted by middle-aged adults and their peers.
    c. are most difficult for women in traditional marriages to adjust to.
    d. are less likely to occur among highly educated adults.

18. According to the text, which does NOT affect parents’ adjustment to the launching phase of the family life cycle (making it rewarding or distressing)? (535-536)
    a. nonparental relationships and roles
    b. parents’ marital and economic circumstances
    c. parents’ educational level
    d. children’s characteristics

19. According to the text, which is NOT a generally mentioned gratification associated with grandparenthood? (537)
    a. being perceived as a wise, helpful person
    b. being able to indulge in grandchildren financially
    c. immortality through descendants
    d. having fun with children without the responsibilities of parenthood

20. Who typically has the closest relationship? (537)
    a. grandmother and granddaughter
    b. grandfather and grandson
    c. grandfather and granddaughter
    d. grandmother and grandson

21. Adult children of aging parents (538)
    a. are not as devoted as past generations were.
    b. spend less time in close proximity to their parents than in previous generations.
    c. generally move farther away from their parents, geographically, as they age.
    d. become less appreciative of their parents strengths and generosity.

22. The number of friends during middle adulthood (542)
    a. increases for men.
    b. increases for women.
    c. decreases for both men and women.
    d. both a and b

23. The glass ceiling (545)
   a. is an invisible barrier that prevents the advancement of women and minorities in corporations.
   b. is something that feminists created in their own minds.
   c. exists because there is less access to mentors, role models, and informal networks that serve as training for women and minorities.
   d. both a and c

24. Midlife career changes (546-547)
   a. are common.
   b. are usually radical.
   c. involve difficult decisions.
   d. typically involve a change from one job to an unrelated line of work.

25. Currently, retirement in America (547-548)
   a. occurs at an older age now than in the past.
   b. is a privilege reserved for the wealthy.
   c. occurs on average at the age of 72.
   d. occurs before the age of 55 one-third of the time.

CHAPTER 17

PHYSICAL AND COGNITIVE DEVELOPMENT IN LATE ADULTHOOD

BRIEF CHAPTER SUMMARY

During late adulthood, individual differences in physical capacities are greater than at any other time in life. A complex combination of genetic and environmental factors combine to determine longevity. More rapid declines occur in the functioning of the brain and autonomic nervous system, sensory systems, cardiovascular and respiratory systems, and immune system, and age becomes more outwardly apparent. Despite these declines, older adults are optimistic about their health. A healthy diet, appropriate exercise and an active stimulating lifestyle continue to contribute to quality and length of life, and sexual enjoyment can be maintained well into late adulthood. Still, many illnesses become more common as age advances, and dementia rises in old age. The most common form of dementia is Alzheimer's disease. In the United States, Alzheimer's is a leading cause of death in late adulthood.

While cognitive declines outweigh advances in late adulthood, high-functioning elders optimize their capacities by becoming more selective in their pursuits and finding new ways to compensate for declines. Memory declines primarily in deliberate processing and prospective memory. These declines affect language production, and hearing loss contributes to difficulties in receptive language. While problem solving becomes less effective, older adults are adept at making quick health-related decisions. And although wisdom is predicted more by life experience than by age, active, respected elders often demonstrate it.

As in earlier age periods, an active mental life predicts maintenance of mental abilities well into old age. Health status also becomes a predictor of cognitive functioning. Terminal decline, or a steady, marked decline in cognitive functioning, indicates that death will soon occur. Older adults benefit greatly from training in cognitive skills and more attention is being given to their continuing education. Many types of programs exist, combining education and meaningful experiences to enrich the lives of senior citizens.

LEARNING OBJECTIVES

After reading this chapter, you should be able to:

17.1. Discuss aging and longevity among older adults. (556-558)

17.2. Describe changes in the nervous system and the senses in late adulthood. (558-563)

17.3. Describe cardiovascular, respiratory, and immune system changes in late adulthood. (563-564)

17.4. Discuss sleep difficulties in late adulthood. (564)

17.5. Describe changes in physical appearance and mobility in late adulthood. (564-566)

17.6. Discuss health and fitness in late life, paying special attention to nutrition, exercise, and sexuality. (566-569)

17.7. Discuss common physical disabilities in late adulthood. (569-576)

17.8. Discuss common mental disabilities in late adulthood, including Alzheimer's disease, cerebrovascular dementia, and misdiagnosed and reversible dementia. (572-576)

17.9. Discuss health care issues that affect older adults, including health care costs and the need for long-term care. (576-578)

17.10. Describe changes in cognitive functioning in late adulthood, including memory, language processing, and problem solving. (579-583)

17.11. Discuss experiences that foster the development of wisdom. (583-584)

17.12. List factors related to cognitive change in late adulthood. (584)

17.13. Discuss the effectiveness of cognitive interventions in late adulthood. (584-585)

17.14. Discuss types of programs and benefits of continuing education in late life. (585-586)

STUDY QUESTIONS

Physical Development

Longevity

1. What is the average human lifespan, under the best of circumstances? What is the maximum human lifespan? (557)
A. ________________________________ B. ________________________________

2. The number of people 65 and older in the United States has risen from _______ million in 1900 to ______ million in 1998. The fastest growing segment of senior citizens is the _____ and older group. (557)

3. List two reasons why the percentage of elderly in the United States is lower than in many other industrialized nations. (557)
A. ____
B. ____

4. True or False: With age, the gender gap in life expectancy narrows, and the gap between whites and ethnic minorities reverses around age 85. (557)

5. Provide evidence of the role of genetics in longevity. (558)
A. ____
B. ____
C. ____

6. List several environmental factors which predict longevity. (558)

____

____

____

7. Despite great diversity, centenarians have many characteristics in common. Summarize several of their common physical and health characteristics. (560-561)

____

____

____

8. In personality, centenarians appear unusually ________________ and ________________, refuse to dwell on ________________ and ________________, are assertive, mention long, happy ________________, and have a history of ________________________ involvement. (560-561)

Physical Changes

1. Neuron loss occurs at different rates in different regions. In the ______________, ______________, and ______________ areas, as many as 50 percent of neurons die. The ________________ (which controls balance and coordination) loses about 25 percent of its neurons. However, parts of the ________________ responsible for integration of information, judgment, and reflection show far less change. (558-559)

2. Several studies indicate that growth of neural fibers takes place at (a slower/the same/a faster) rate in older, healthy adults in comparison to middle-aged adults. ________________________ can help preserve brain structures and behavioral capacities. (559)

3. List three changes in autonomic nervous system functioning with age. (555)

A. ____

B. ____

C. ____

4. In late adulthood, vision continues to decline as the cornea becomes more ________________________, the lens continues to ______________, and ________________________ develop. Are cataracts treatable? ________ (559)

5. In addition to blurred, glare-sensitive, foggier vision resulting from the changes described in question 4, dark ________________, ________________ perception, and visual ________________________ decline. (560)

6. What is the leading cause of blindness among older adults? (560)

____

7. True or False: Visual difficulties have little impact on older people's self-confidence or every day behavior. (561)

8. Hearing impairments are far (more/less) common than vision problems, and more (men/women) are affected. (562)

9. Of all hearing difficulties, the decline in ______________ perception has the greatest impact on life satisfaction. (562)

10. As with vision, most elders (do/do not) suffer from hearing loss severe enough to impair daily functioning. What can family members do to help older people demonstrate their alertness and competence in conversation? (562)

____

11. True or False: While reduced taste sensitivity is probably due to other factors, reduced smell is due to age-related declines. (562)

12. What two factors are thought to contribute to reduced touch perception in the elderly? (562)
A. ____
B. ____

13. Three important changes take place in the heart over time. Due to increased rigidity, death, and enlargement of cells, the walls of the ______________________________ thicken. Artery walls ________________ and accumulate ________________. Finally, the heart muscle becomes less responsive to ____________________ cells. (563)

14. Due to reduced __________________, the vital capacity of the lungs is reduced by half. (563)

15. ________________ system functioning becomes less effective, increasing risk of illness, and is more likely to malfunction by turning against the body in an ______________________ response. (563)

16. Immune functioning in old age varies (little/greatly) and is a sign of ____________________________________. (563)

17. Impaired immune functioning is determined by both genetics and other physical changes of aging such as increased production of ______________ hormones. List two factors which can help protect the immune response. (563-564)
A. ______________________________ B. ______________________________

18. Changes in ______________________________ controlling sleep and higher levels of ____________________________ in the bloodstream are believed to be responsible for increased sleep difficulties in late adulthood. (564)

19. List three reasons why men experience more sleep difficulties than women until the age of 70 or 80. (564)
A. ____
B. ____
C. ____

20. True or False: Prescription sedatives provide a beneficial, long-term solution to the sleep difficulties of the elderly. (564)

21. Describe several changes in physical appearance which occur during late adulthood. (564-565)

____
____
____
____
____

22. Mobility declines as ________________ and ________________ strength decline faster than at earlier ages, as does the flexibility and strength of ________________ and ________________. (565)

23. Certain ________________ and ________________ exercises, such as dance, can be helpful in maintaining flexibility and range of movement. (565)

24. People who avoid confronting age-related change, viewing it as inevitable and uncontrollable, report (more/less) physical and psychological changes. (565)

25. The more older people subscribe to ________________________ and are treated in ________________________ ways, the less likely they are to cope adaptively with age-related change. (565)

26. In what kind of cultures is coping among the elderly much easier? (565-566)

____

27. True or False: Among the Herero of Botswana, being sent to provide live-in care for frail elders is a source of great pride and prestige. (567)

28. In the United States, communities with highly (stable/unstable) populations may foster more inclusion of the elderly in community life. (567)

Health, Fitness, and Disability

1. True or False: Although their number of feared possible physical selves increases with age, older adults generally rate their physical health as favorably as do college students. (566)

2. What two aspects of health are intimately related in late life? (566)
A. ______________________________ B. ______________________________

3. Before age 85, ethnic minority groups are at (greater/less) risk for certain health problems. Why are they less likely to seek and comply with treatment? (567-568)

____

4. True or False: By age 85 and beyond, women are more impaired than men because only the strongest men have survived. (568)

5. As life expectancy increases, we want the average period of ________________ to decline. This is called the ______________________________. (568)

6. List two impairments which increase the risk of dietary deficiencies. (568)
A. ____
B. ____

7. True or False: Vitamin-mineral supplements have little effect on health and physical functioning. (568)

8. True or False: Endurance training and weight-bearing exercise only have important health benefits when begun by the end of middle adulthood. (568)

9. Although almost all cross-sectional studies report a decline in sexual desire and activity, the trend may be exaggerated by ______________________. Most healthy older married couples report (declining/continued) sexual enjoyment. (569)

10. (Men/Women) are more likely to cease to interact sexually. (569)

11. List factors responsible for reduced sexuality in older men and women. (569)
Men: ____
____
Women: ____

12. List the six leading causes of death for those 65 and older. (569-570)
A. ______________________________ B. ______________________________
C. ______________________________ D. ______________________________
E. ______________________________ F. ______________________________

13. True or False: Physical and mental disabilities that are *related* to aging are *caused* by aging. (570)

14. ______________________ is the most common form of arthritis, is related to years of use, and is limited to certain joints. ____________________ arthritis involves the whole body, is often severe, and is caused by an autoimmune response. (570)

15. List three components of arthritis management. (570)
A. ______________________________ B. ______________________________
C. ____

16. Adult-onset diabetes occurs when ____________ levels are no longer able to control the amount of _______________ in the bloodstream, causing a variety of health problems. (570-571)

17. What factors are related to the risk of developing adult-onset diabetes? Circle the letters of the two which are a focus of change in treatment. (571)
A. ______________________________ B. ______________________________
C. ______________________________

18. True or False: Unintentional injuries are at an all-time high in late adulthood. (571)

19. True or False: Per mile driven, rates of traffic violations, accidents, and fatalities for older adults are higher than for any other group except those under age 25. (571)

20. List three characteristics which contribute to the driving difficulties of older adults. (571)
A. ______________________________ B. ______________________________
C. ______________________________

21. Nearly _____ percent of deaths between ages 65 and 74 in the U.S. result from pedestrian accidents. (571)

22. One of the most common injuries resulting from falls is ________ fracture. Of those who survive, only ___________ regain the ability to walk without assistance. (571)

23. How can falling impair health indirectly? (572)
____

24. Training that enhances _______________ and _______________ skills necessary for safe driving and that helps older adults avoid _______________ situations while driving can save lives. (572)

25. List three ways to help prevent falls among the elderly. (572)
A. _______________ B. _______________
C. _______________

26. The most common form of dementia, also a leading cause of death in late adulthood, is _______________ disease. (573)

27. Severe _______________ problems are often the earliest symptom of Alzheimer's, followed by faulty _______________, _______________ changes, and _______________. (573)

28. As Alzheimer's progresses, _______________ movements disintegrate, sleep is disrupted by _______________, the ability to comprehend and produce _______________ is lost, and finally the ability to _______________ familiar people deteriorates. (573)

29. Identify two major structural changes and one type of chemical change that appear more pronounced in the brains of Alzheimer's patients. (573)
A. _______________
B. _______________
C. _______________

30. Briefly describe the two types of Alzheimer's disease. (573-574)
A. _______________
B. _______________

31. The most consistent risk factor for Alzheimer's disease is _______________, but this risk factor is not found in 50 percent of cases. (574)

32. List three factors which may help protect against Alzheimer's. (574)
A. _______________ B. _______________
C. _______________

33. _______________ and _______________ drugs may be prescribed to help control the behavior of Alzheimer's victims, while _______________ and social _______________ can help caregivers respond with patience and compassion. Dramatic changes in _______________ should be avoided to promote a feeling of security. (574)

34. List factors associated with cerebrovascular dementia. (575)
____
____
____
____

35. In most cases, cerebrovascular dementia is caused by _______________. List the warning signs of stroke. (575)
____
____
____

36. The disorder most often misdiagnosed as dementia is ______________. List three additional factors which can lead to reversible symptoms of dementia. (575-576)
A. ________________________ B. ________________________
C. ________________________

37. Americans age 65 and older account for ________ percent of the population, but ________ percent of federal health care spending. This cost is expected to triple by the year ________. (577)

38. Summarize three promising approaches to freeing up public resources in the area of health care for the elderly. (576)
A. ________________________
B. ________________________
C. ________________________

39. What two disorders of aging most often lead to nursing home placement? (578)
A. ________________________ B. ________________________

40. Since funding for nursing home care is not generally provided until resources are exhausted, the users of such care are most often people with ______________ and ______________ incomes. (578)

41. Which group is most likely to utilize nursing home care? (578)
A. African Americans B. Asian Americans C. Caucasian Americans

42. List three factors that predict life satisfaction among institutionalized elderly. (579)
A. ________________________ B. ________________________
C. ________________________

Cognitive Development

1. True or False: While losses outweigh gains, individual differences in cognitive functioning are greater in late adulthood than during any other time period. (579)

2. Elders who sustain high functioning often ______________ their goals in order to ______________ returns on their diminishing energy, and ______________ for losses in various ways. (579)

Memory

1. Memory problems are especially evident on ______________ tasks. (580)

2. ______________ cues serve as important retrieval cues when remembering. (580)

3. ______________ memory suffers far less than ______________, suggesting that providing more memory ________ would enhance older adults' performance. (580)

4. Age-related memory declines are largely limited to tasks requiring ______________ processing rather than those which occur automatically, such as ______________, or nonconscious, memory. (580)

5. True or False: Older adults' remote, or very long-term recall, is typically much clearer than their memory for recent events. (581)

6. Adults age 50 to 90 recall both remote and recent events more easily than intermediate events. Explain why. (581)

____

____

7. Why do adults do better on *event-based* than *time-based* prospective memory tasks? (581-582)

____

____

8. Explain why older people's difficulty with prospective memory does not always show up in real life. (582)

____

Language Processing

1. Like implicit memory, ______________________ changes very little in late life. (582)

2. What two aspects of language production become more difficult with advancing age? (582)

A. ____

B. ____

3. List two ways older adults compensate for difficulties in language production. (582)

A. ____

B. ____

Problem Solving

1. Problem solving (improves/declines) in late life as older adults seem less willing to ______________________ when a problem calls for it and because ____________ limitations make it hard to attend to all relevant facts. (583)

2. In making decisions about seeking medical care, older adults decide (quicker/slower) than do younger adults. (583)

Wisdom

1. List capacities most people mention in their descriptions of wisdom. (583)

A. ____

B. ____

C. ____

D. ____

2. Type of ______________________ appears to have more of an impact on the development of wisdom than does age. People in ________________ careers are more likely to make wise decisions, but so are active, highly regarded ________________ citizens from all walks of life. In addition, overcoming ____________________ also appears to contribute to late-life wisdom. (583-584)

Factors Related to Cognitive Change

1. List four factors that predict maintenance of mental abilities in late life. (584)
A. ______________________ B. ______________________
C. ______________________ D. ______________________

2. Retirement affects cognitive development positively when people leave routine jobs in favor of ______________________, but not when they leave highly ____________ jobs without developing challenging substitutes. (584)

3. The steady, marked decline in cognitive functioning, called ____________ ____________, on average, lasts about ____ years. (584)

Cognitive Interventions

1. When provided cognitive training, ____ percent of individuals over age 64 who had shown previous declines returned to the level at which they had been functioning ____ years earlier. (584)

2. List two other areas in which the elderly have been shown to improve as a result of training. (584)
A. ______________________ B. ______________________

Lifelong Learning

1. Describe what Elderhostel programs provide. (585)
____
____
____

2. The University of the Third Age, originating in France, offers various ______________________ sponsored courses for the elderly, and in Australia and Great Britain, ____________ often do the teaching. (585)

3. Some programs foster ______________________ relationships and ______________________ involvement, teaching grandparents skills needed to help grandchildren with school work, for example. Returning to ____________ is another option for seniors. (585)

4. List characteristics of those who are most likely to attend learning programs for the elderly. (585)
A. ______________________ B. ______________________
C. ______________________

5. List benefits of participation in continuing education. (586)
____
____
____
____

ASK YOURSELF . . .

Sixty-five-year-old Herman inspected his thinning hair and bare scalp in the mirror. "The best way to adjust to this is to learn to like it," he thought to himself. "I remember reading somewhere that bald older men are regarded as leaders." What type of coping is Herman using, and why is it an effective way to adapt to this aspect of physical aging? (see text pages 565-566)

Reread the story of Dr. Stojic, a robust centenarian, in the Lifespan Vista box on page 560. What aspects of his life history are consistent with research findings on factors that contribute to a long and healthy life? (see text pages 556-558)

Marissa complained to a counselor that at age 68, her husband, Wendell, stopped initiating sex and no longer touched, stroked, or cuddled her. Why might Wendell have ceased to interact sexually? What interventions—both medical and educational—could be helpful to Marissa and Wendell? (see text pages 568-569)

Explain how depression can combine with physical illness and disability to promote cognitive deterioration in the elderly. Should cognitive declines due to physical limitations and depression be called dementia? Why or why not? (see text pages 575-576)

When Ruth couldn't recall which movie Walt was thinking about, she asked several questions: "Which theater was it at, and who'd we go with? Tell me more about the little boy [in the movie]." Which memory deficits of aging is Ruth trying to overcome? (see text pages 580-582)

Ruth complained to her doctor that she had trouble finding the right words to explain to a delivery service how to get to her house. What cognitive changes account for Ruth's difficulty? (see text page 582)

Describe cognitive functions that are maintained and that improve in late adulthood. What aspects of aging contribute to them? (see text pages 583-584)

SUGGESTED READINGS

Craik, F., & Salthouse, T. A. (Eds.). (1999). *The handbook of aging and cognition* (2nd ed.). Mahwah, NJ: Erlbaum. Examines a broad range of topics dealing with developmental trends in cognition, including age-related changes in perception, memory, language, emotion, motivation, and personality.

Manuck, S. B., Jennings, R., Rabin, B., & Baum, A. S. (Eds.). (1999). *Behavior, health, and aging*. Mahwah, NJ: Erlbaum. Discusses the impact of the increasing number of elderly Americans on the health care system. Addresses issues relevant to basic changes that accompany aging, including management of chronic disease, menopause, and effects of aging on the immune system.

Rogers, W. A., Fisk, A. D., & Walker, N. (1996). *Aging and skilled performance: Advances in theory and application*. Mahwah, NJ: Erlbaum. Discusses whether, how, when, and why a variety of cognitive skills (e.g., speed of acquisition, accuracy of performance, and retention over time) in a wide range of domains declines with age. Explores applications of research findings to enhancing or enabling the daily living activities of elderly adults.

Whitman, T. L., Merluzzi, T. V., & White, R. D. (Eds.). (1998). *Life-span perspectives on health and aging*. Mahwah, NJ: Erlbaum. Provides an overview of the changing biological, psychological, and environmental influences on health from the prenatal period through late adulthood. Examines the development of major biological systems and the changing role of genetics and environment over time.

PUZZLE 17.1 TERM REVIEW

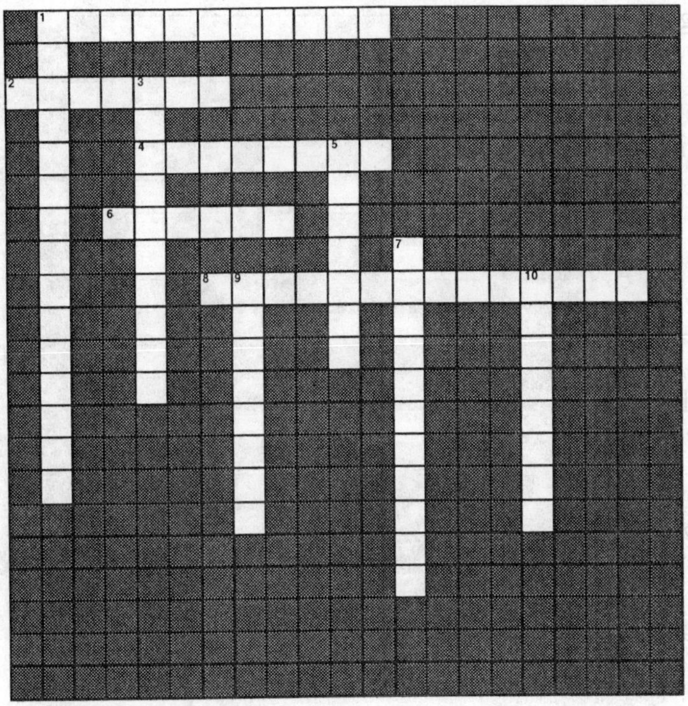

Across

1 _____ of morbidity: public health goal of reducing the average period of diminished vigor before death as life expectancy extends

2 _____ aging: genetically influenced age-related declines in functioning that affect all members of our species and take place even in the context of good health

4 _____ decline: steady, marked decrease in cognitive functioning prior to death

6 Form of cognition combining breadth and depth of practical knowledge, ability to reflect on that knowledge in ways that make life more worthwhile, emotional maturity, and creative integration of experience and knowledge into new ways of thinking and acting

8 Form of arthritis in which cartilage in frequently used joints deteriorates, leading to swelling, stiffness, and loss of flexibility

Down

1 In _____ dementia, a series of strokes leaves dead brain cells, producing step-by-step degeneration of mental ability.

3 _____ response: abnormal response in which the immune system turns against normal body tissues

5 _____ plaques: structural change in the brain associated with Alzheimer's disease in which deposits of the protein are surrounded by clumps of dead nerve cells, protein appears to destroy

7 _____ memory: recall that involves remembering to engage in planned actions at an appropriate time in the future

9 _____ of the population pyramid: increasing percentage of older adults and decreasing percentage of younger adults in the population due to declining birth rates and greater longevity

10 Intergenerational _____: circumstance in which the elderly receive more than their fair share of public resources at the expense of the very young, due to their numbers and ability to vote and lobby for their needs

PUZZLE 17.2 TERM REVIEW

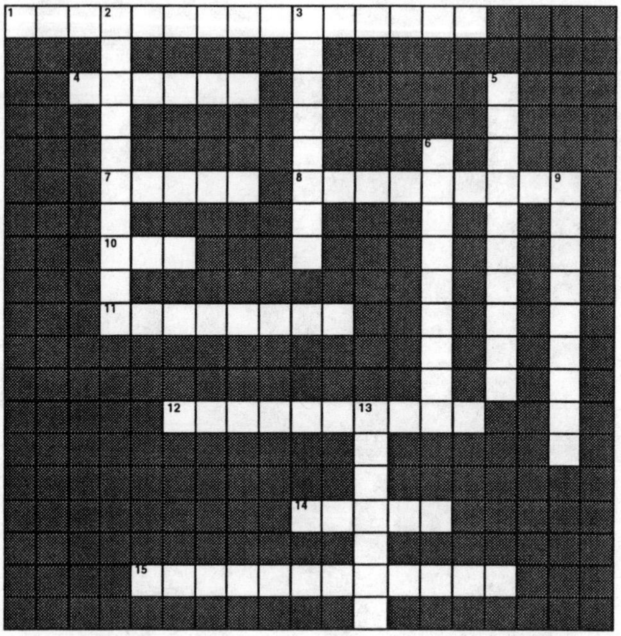

Across

1 _____ tangles: bundles of twisted threads within neurons accompanying Alzheimer's disease
4 ____ memory, or very long-term recall
7 Sleep _____, or cessation of breathing for at least 10 seconds
8 Cloudy areas in the lens of the eye, resulting in foggy vision
10 _____-old, or elderly who show signs of decline
11 A set of disorders in which many aspects of thought and behavior are impaired
12 _____ disease, the most common form of dementia
14 _____-old, or elderly who appear physically young for their advanced years
15 Selective _____, ways of maximizing returns from diminishing energy

Down

2 _____ arthritis, or inflammation of connective tissue due to an autoimmune response
3 Memory without conscious awareness
5 _____ age, or actual competence and performance
6 Life expectancy _____, when surviving members of ethnic minority groups live longer than members of the white majority
9 A type of aging in which declines are due to hereditary defects and negative environmental influences
13 _____ degeneration, caused by restriction of blood flow to the central region of the retina

SELF-TEST

1. The rate of aging (556)
   a. is constant for most people.
   b. can be predicted by scientists based on biological measures.
   c. is expressed in terms of functional age (young-old and old-old).
   d. is determined by chronological age.

2. The number of Americans age 85 and older (557)
   a. has decreased since 1900 largely because of pollution.
   b. currently represent less than 2 percent of the population.
   c. is expected to quadruple by 2050.
   d. is both b and c.

3. The largest percentage of what group is most likely to live to be 90? (558)
   a. men
   b. women
   c. individuals whose parents both live to be 70 or older
   d. ethnic minorities (African Americans, Hispanics, or Native Americans)

4. Which is NOT the result of normal deterioration of the nervous system in old age? (558-559)
   a. loss of balance and coordination.
   b. heart attacks
   c. lower tolerance for extremes of hot and cold weather
   d. difficulty remembering

5. Which of the following is the leading cause of blindness among older adults? (560)
   a. cataracts
   b. yellowing of the lens
   c. cell loss in the retina and optic nerve
   d. macular degeneration

6. Among people age 85 and older, ____ percent experience visual impairment severe enough to interfere with daily living. (561)
   a. 25
   b. 30
   c. 40
   d. 50

7. Which slows the effects of aging on the cardiovascular system? (563)
   a. exposure to environmental pollutants
   b. smoking
   c. excess dietary fat
   d. exercise

8. Which is NOT related to the normal decline in immune functioning in old age? (563)
   a. cardiovascular disease
   b. AIDS
   c. rheumatoid arthritis
   d. diabetes

9. Older adults' difficulties related to sleep (564)
   a. are only because they have trouble falling asleep.
   b. are only because they have trouble staying asleep.
   c. are due to older adults' need for less sleep than younger adults.
   d. begin after the age of 30 for men and after age 50 for women.

10. Which is NOT a common change in physical appearance during old age? (564-565)
    a. skin imperfections such as moles, age spots, wrinkles, and dryness
    b. thinning hair
    c. thinning bodies and thickening arms and legs
    d. broadening of the nose and ears

11. Which factor does NOT commonly affect mobility in late adulthood? (565)
    a. loss of muscle power
    b. rheumatoid arthritis
    c. deterioration of bone strength
    d. loss of flexibility in tendons and ligaments

12. Older adults are generally (566)
    a. optimistic about their health.
    b. pessimistic about their health.
    c. indifferent about their health.
    d. helpless and hopeless concerning their health.

13. Research indicates that by taking vitamin-mineral supplements, older adults (568)
    a. can reduce the number of days they spend sick by nearly 50 percent.
    b. can lower their risk of cardiovascular disease.
    c. can help prevent macular degeneration.
    d. can do both a and c.

14. Sedentary adults who begin endurance training in their seventies (568)
    a. will show gains in vital capacity similar to those of younger adults.
    b. are wasting their time.
    c. should not engage in frequent, strenuous exercise. Low-intensity exercise is sufficient to increase the heart's pumping capacity.
    d. experience both a and c.

15. Older adults' sexual activity (568-569)
    a. ends for the most part by age 60.
    b. declines drastically because of major changes in the reproductive organs.
    c. continues to be regularly enjoyed by healthy individuals until death.
    d. is both a and b.

16. Which is NOT a physical disability common in old age? (569-570)
    a. stroke
    b. osteoarthritis
    c. adult-onset diabetes
    d. leukemia

17. Which is NOT true concerning Alzheimer's disease? (573-574)
    a. It is the most common form of dementia.
    b. It is found in 5 to 7 percent of adults over the age of 65.
    c. It only affects individual's ability to remember.
    d. It is caused by structural changes in the brain and lower levels of neurotransmitters necessary for communication between neurons.

18. Which of the following is NOT true about dementia? (575-576)
    a. Most types are curable.
    b. Alzhemier's is the most common form.
    c. Memory problems are often the earliest symptom.
    d. Personality changes and depression often appear.

19. Which is NOT a health care issue that particularly affects older adults? (576-578)
    a. government funding for healthcare
    b. the fact that 5 percent of Americans age 65 and older are in nursing homes
    c. publicly funded in-home healthcare
    d. long-term healthcare choices

20. Which type of memory function is likely to decline the LEAST in old age? (580-582)
    a. recognition
    b. remote memory
    c. recall
    d. prospective memory

21. According to the text, older individuals are less willing to seek further information when a problem calls for it, except when the problem pertains to their (583)
    a. career.
    b. family.
    c. health.
    d. home.

22. Wisdom (583)
    a. can only be developed through a long lifetime of experiences.
    b. is gained by all older adults.
    c. is gained by only older adults.
    d. is demonstrated by a very small number of adults from all age groups.

23. Which is NOT a factor related to cognitive change in late adulthood? (584)
    a. above-average education
    b. community involvement
    c. a flexible personality
    d. number and ages of children

24. Cognitive skills training (584)
    a. can reduce the rate of intellectual decline in older adults, but only if the training is given early in life (before age 50).
    b. is of no use to older adults whose intellectual declines are due to disuse of particular skills.
    c. lasting only 5 hours may effectively improve cognitive skills among elderly adults.
    d. may improve elderly adults' mental functioning, but any improvements are temporary. Within 6 weeks, any gain made will no longer exist.

25. What is an Elderhostel program? (585)
    a. a type of nursing home
    b. 1- to 3-week long college courses designed for the elderly
    c. a place where old people can stay for free while traveling in Europe
    d. a German program that provides vocational training for the elderly

CHAPTER 18

EMOTIONAL AND SOCIAL DEVELOPMENT IN LATE ADULTHOOD

BRIEF CHAPTER SUMMARY

Erikson's psychological conflict of ego integrity versus despair is faced in late adulthood. Those with a sense of integrity accept their life course with its imperfections, while those who experience despair feel they have too little time to overcome disappointing failures. According to Peck, integrity requires that adults move beyond their work, their bodies, and their separate identities and work to make the world better. Labouvie-Vief believes integrity is facilitated by adults' fuller understanding of their emotions. Current research indicates that elderly people engage in a form of reminiscence called life review as part of achieving ego integrity, preventing despair, and accepting the end of life. Older adults' personalities become more agreeable, less extroverted, and more accepting of change, and spirituality may advance to a higher level.

Continuing social changes begun in middle adulthood, elders choose their social circle more selectively, focusing on a few familiar, pleasurable relationships. Most remain in the homes where they spent their adult lives, with increasing numbers living alone. While restriction of autonomy in U.S. nursing homes leads to social withdrawal, residential communities often increase opportunities for meaningful interaction.

Marital happiness is at its peak in late adulthood. Although divorce is more stressful in late life, it is very rare, and remarriages are more successful than at younger ages. Widowhood is highly stressful, but older adults adjust better than those widowed at younger ages. Never-married elders, mostly women, develop alternative ties and generally fare well. Siblings, friends, and adult children provide important sources of emotional support and companionship to elders. Relationships with grandchildren and great-grandchildren provide meaningful links to the future. However, some elders are maltreated, a situation affected by various characteristics of perpetrators, victims, and institutions.

The decision to retire involves many factors, and while most elders adjust well, reluctance to retire predicts stress. Senior leisure activities are typically an extension of those engaged in earlier in life. Volunteer work and increased political activism offer important opportunities for elders to contribute to society. Successful aging, involving maximizing gains and minimizing losses, is best viewed as a process rather than a list of accomplishments and is facilitated by cultural policies that value the contributions of older adults and support their continued development.

LEARNING OBJECTIVES

After reading this chapter, you should be able to:

18.1. Describe Erikson's stage of ego integrity versus despair. (592-593)

18.2. Describe Peck's and Labouvie-Vief's views of development in late adulthood, and discuss the functions of reminiscence and life review in older adults' lives. (593-594)

18.3. Describe stability and change in self-concept and personality in late adulthood. (595)

18.4. Discuss spirituality and religiosity in late adulthood, and trace the development of faith. (596-599)

18.5. Discuss individual differences in psychological well-being as older adults respond to challenges posed by issues of control versus dependency, declining health, and negative life changes. (599-602)

18.6. Describe the role of social support and social interaction in promoting physical health and psychological well-being in late adulthood. (602)

18.7. Describe social theories of aging, including disengagement theory, activity theory, and socioemotional selectivity theory. (603-604)

18.8. Discuss the impact of communities, neighborhoods, and fear of crime on elders' social lives. (604-606)

18.9. Discuss the effects of different housing arrangements on older adults' adjustment. (606-608)

18.10. Describe changes in social relationships in late adulthood, including marriage, divorce, remarriage, and widowhood, and discuss social experiences and life satisfaction of never-married, childless older adults. (609-612)

18.11. Describe late-life sibling relationships and friendships. (612-613)

18.12. Describe older adults' relationships with adult children, adult grandchildren, and great-grandchildren. (614-615)

18.13. Discuss elder maltreatment, including risk factors and strategies for prevention. (615-616)

18.14. Discuss retirement and leisure, paying special attention to the decision to retire, adjustment to retirement, and leisure activities. (616-619)

18.15. Discuss the meaning of successful aging. (619-620)

STUDY QUESTIONS

Erikson's Theory: Ego Integrity versus Despair

1. Describe the feelings of adults who arrive at a sense of ________________, the positive outcome of Erikson's final conflict. (592)

____

2. The capacity to view one's own life in the larger context of ______________________ contributes to the serenity and contentment of integrity. (592)

3. The negative outcome of this stage is ______________, and occurs when elders feel they have made ______________________________ and time is too short to find an alternative life path. How did Erikson believe these attitudes are often expressed? (593)
____

Other Theories of Psychosocial Development in Late Adulthood

1. List the three tasks Peck viewed as comprising Erikson's conflict of ego integrity versus despair. (593)
A. ____
B. ____
C. ____

2. Labouvie-Vief discovered that with maturity, adults arrive at a fuller understanding of ______________________ and are better able to interpret ______________ events in a positive light. (593)

3. According to Butler, most older adults engage in ____________________, a special kind of reminiscence, as part of attaining ego integrity, preventing despair, and accepting the end of life. (594)

4. True or False: Findings suggest that older adults wish to be young again. (594)

5. List three factors which color elders' recollections of distant events. (594)
A. ______________________________ B. ______________________________
C. ______________________________

6. List three reasons the elderly reminisce, in addition to life review. (594)
A. ______________________________ B. ______________________________
C. ______________________________

7. Separation from parents and inadequate care had a more lasting impact on temperamentally ______________, ________________ evacuees and refugees than on outgoing ones. Adolescents fared best when placed in ______________________, while children were better off when placed in ______________________. (596-597)

8. Rather than healing with the passage of time, some unresolved childhood traumas can ______________ in later life. Nevertheless, (many/very few) World War II interviewees were maladjusted in later life. (596-597)

Stability and Change in Self-Concept and Personality

1. Because of a lifetime of self-knowledge, older adults' conceptions of themselves are more ________________ and ________________. They are more likely to show self-________________. (595)

2. True or False: Most elderly people no longer mention hoped-for selves. (595)

3. Identify three frequent personality changes which occur in old age. (595)
A. ______________________________ B. ______________________________
C. ______________________________

4. Older adults attach (great/moderate/little) value to religious beliefs and behaviors. (596)

5. True or False: Involvement in religion remains fairly stable throughout adulthood. (597)

6. What three changes occur in adult faith development between Fowler's fourth and fifth stages? (597-598)
A. ______________________________
B. ______________________________
C. ______________________________

7. Organized and informal religious activities are especially high among ____________ ______________________ elders. (598)

8. List some positive outcomes of religious involvement. (598-599)
____
____

9. (Men/Women) are more likely to be involved in religion. What might explain this trend? (599)
____

Individual Differences in Psychological Well-Being

1. The ______________________ script and ______________________ script reinforce dependent behavior at the expense of independent behavior. (599)

2. What factors influence whether elders will react positively or negatively to caregiving? (599)
____
____

3. Dependency can be adaptive if it permits older people to ______________________ ______________________. (600)

4. Physical illness can lead to a sense of loss of personal ______________ and is among the strongest risk factors for late-life ______________. (600)

5. What occurs for many elders within the first month of nursing home residence? (601)
____

6. Suicide reaches its highest rate among people age _____ and older. What explains the much lower rate of suicide among women and ethnic minorities? (600-601)
____
____

7. List two reasons why elder suicides tend to be underreported. (600-601)
A. ______________________________
B. ______________________________

8. What two types of events often prompt suicide in late life? (600-601)
A. ____
B. ____

9. List warning signs of elder suicide. (601)
____
____

10. The most effective treatments for suicidal elders who are depressed combine ____________________ with ____________________. (601)

11. True or False: Negative life changes may cause less stress for older people than for younger ones. (602)

12. When negative changes pile up, ____________________ continues to play an important role in reducing stress and promoting well-being. (602)

13. (Men/Women) of advanced age report a lower sense of psychological well-being than do (men/women). (602)

14. Formal support can complement informal assistance when elders do not want family support they cannot ____________________. (602)

15. List two factors which make ethnic minority elders more likely to accept formal support. (602)
A. ____
B. ____

16. (Extroverted/Introverted) elders are more likely to maintain high morale. (602)

A Changing Social World

1. According to ____________________ theory, mutual withdrawal between elders and society takes place in anticipation of death. (603)

2. Provide evidence suggesting that the above theory does not adequately explain the reduced social activity of older people. (603)
____

3. ____________________ theory states that social barriers to engagement, as opposed to the desires of elders, cause declining rates of interaction. (603)

4. Provide evidence that the above theory does not account for changes in social interaction among the elderly. (603-604)
____
____

5. According to socioemotional selectivity theory, social interaction declines with age due to the changing functions of social interaction. List two functions which decline in old age and one which elders emphasize. (604)
Declines: A. ____________________ B. ____________________
Emphasis: ____

6. True or False: When younger people are asked who they would spend their time with if they would soon be leaving their community, they express a preference for familiar partners, just as older people do. (604-605)

7. List three types of communities from highest to lowest in terms of senior citizens' incomes, health, and access to social services. (604, 606)
A. ______________________________ B. ______________________________
C. ______________________________

8. Older adults report greater life satisfaction when ______________________ reside in their neighborhoods. (606)

9. Fear of ______________________ has a serious negative impact on older adults' sense of security and comfort. (606)

10. True or False: The elderly are less often victims of violent crime than are people of other ages, but are more likely to be victims of purse-snatching and pick pocketing. (606)

11. ______ percent of older adults choose to stay in their own homes and neighborhoods. (606-607)

12. True or False: Most ethnic minority elders would prefer to live with their children. (607)

13. Currently, ______________ of older Americans live alone, with ______ percent of those age 85 and older doing so. Over _____ percent of them are poverty stricken, and most are ______________________. (607)

14. True or False: Elderly women living alone in other Western nations face the same dire circumstances as those in the United States. (607)

15. Identify three types of residential communities for older people. (608)
A. ______________________________ B. ______________________________
C. ______________________________

16. List three positive effects of residential communities on physical and psychological health. (608)
A. ______________________________
B. ______________________________
C. ______________________________

17. List four aspects of residential communities which are important in leading to the above benefits. (608)
A. ______________________________
B. ______________________________
C. ______________________________
D. ______________________________

18. List two reasons why interaction within the nursing home does not predict residents' life satisfaction. (608)
A. ______________________________
B. ______________________________

19. Why have efforts to design more home-like nursing homes such as those in Europe been unsuccessful in the United States? (608)

____

Relationships in Late Adulthood

1. Provide three reasons why marital satisfaction increases from middle to late adulthood, when it is at its peak. (609)
A. ____
B. ____
C. ____

2. When marital dissatisfaction is present, it takes a greater toll on (men/women) who tend to confront and try to solve marital problems rather than retreat. (609)

3. Although rare, divorce in late adulthood is increasing. What reasons do men and women give for initiating divorce in late life? (609)
Men: ____
Women:____

4. In younger adults, divorce often leads to greater awareness and resolve to change ____________________________________. In contrast, ________ __________________ in divorced elders heightens guilt and depression because their self-worth depends more on __________ accomplishments. (609-610)

5. List three reasons why late adulthood remarriages are more frequent after divorce than widowhood. (610)
A. ____
B. ____
C. ____

6. Remarriages that occur in (young/middle/late) adulthood are more likely to be satisfying and less likely to end in divorce. (610)

7. When widowed elders move due to financial or physical difficulties, they usually choose to move (closer to/in with) family. (610)

8. When a spouse dies, _______________ and ___________________ friends often withdraw from the widowed person's support system, but _________________ usually step in. (610)

9. (Men/Women) find it more difficult to adjust to widowhood. What areas do male and female widows mention as their greatest areas of strain and concern? (611)
Men: ____
Women:____

10. In addition to greater availability of potential partners, men are more likely to remarry than are women because men often lack skills for maintaining _________________ relationships, forming satisfying emotional bonds outside of __________________, and handling the __________________ of their deceased wives. (611)

11. Adjustment to widowhood (does/does not) vary across cultures. (611)

12. List five suggestions for fostering adaptation to the loss of a spouse in old age. (611)
A. ________________________________ B. ________________________________
C. ________________________________ D. ________________________________
E. ________________________________

13. Americans who remain single and childless throughout life usually develop alternative ________________________, and in the case of women, ____________________ tend to be unusually close. (612)

14. Never-married elderly women report a level of life satisfaction (higher than/equal to/lower than) that of married elders and (higher than/ equal to/lower than) that of divorcees and the recently widowed. (612)

15. Both men and women perceive bonds with (same-sex siblings/sisters) to be closer than other sibling bonds. The closer the tie to one of these relatives, the higher the elder's ____. (612)

16. Elderly siblings in industrialized nations are more likely to (socialize/ provide direct assistance), but both occur, especially for ____________________ and __________________________________ elders. (612)

17. Because they share a long, unique history, elderly siblings often engage in more joint ____________________. (612)

18. List four functions of elderly friendships. (612-613)
A. ____
B. ____
C. ____
D. ____

19. True or False: Friendship formation continues throughout life, and the elderly report more intergenerational friends. (613)

20. Elderly (men/women) have more intimate friends; Elderly (men/women) rely more on female relatives than friends for open communication; Older (men/women) have more secondary friends. (613)

21. In elder friendships, maintaining a healthy ________________ in affection and emotional support given and received is important. Tensions can arise in elderly friendships when one member tries to ______________ too much. (613)

22. True or False: Older parents expect more emotional support than practical assistance from adult children and usually only seek help from them when there is a pressing need. (614)

23. True or False: Sons are more likely to arrange family contacts for aging parents, and elderly mothers and fathers are equally likely to maintain intimate contact with their adult children. (614)

24. The majority of grandchildren (do/do not) feel obligated to assist aging grandparents. Grandparents (do/do not) regard the adult grandchild tie as very gratifying and as a vital link between ____________________________ and ______________________. (614-615)

25. Although contact between grandparents and grandchildren usually declines over time, grandchildren are becoming increasingly important sources of ____________________ for elders. (615)

26. Forty percent of American elders have great-grandchildren, viewing their new role as (active/limited) and as a sign of ________________________. They (are/are not) generally enthusiastic about it. (615)

27. Elder maltreatment is lower in ethnic groups with strong traditions of ____. (615)

28. List four forms of elder abuse which occur with about __________ percent of all elders. Indicate which are the most common and second most common forms. (615)

A. ____________________________ B. ________________________________

C. ____________________________ D. ________________________________

29. The perpetrator of abuse is usually someone the elderly adult loves, trusts, and depends on. List the most likely perpetrators in order of prevalence. (615)

A. ____________________________ B. ________________________________

C. ____

30. List five risk factors which contribute to the likelihood of elder abuse. (615-616)

A. ____

B. ____

C. ____

D. ____

E. ____

31. Identify several components of elder maltreatment prevention. (616)

____

____

____

32. True or False: Legal action against elder abuse is common. (616)

33. Societal efforts at reducing elder abuse include public education to encourage _______________ of abuse and understanding of the ____________ of older people, as well as countering negative ______________________ of aging. (616)

Retirement and Leisure

1. Almost ___________ of American retirees re-enter the labor force, most within ___________ of initial retirement. (616-617)

2. What is usually the first consideration in the decision to retire? List additional considerations in the decision to retire. (617)

First: ______________________ Additional: ______________________________

____

____

3. On the average, (men/women) retire earlier because of family events, but those who are ________________ are an exception, and often must continue working into old age. In some other Western nations, higher ____________ make retirement feasible for the economically disadvantaged and time devoted to ________________ is given credit when figuring benefits. (617)

4. Adjustment to retirement can be likened, in many cases, to the adjustment of younger people who experience ________________. (618)

5. True or False: Health problems lead elders to retire rather than vice versa, and mental health remains fairly stable for most people from pre- to post-retirement. (618)

6. Factors associated with ______________________________ predict stress following retirement. List four other factors associated with adjustment to retirement. (618)

A. ______________________________ B. ______________________________

C. ______________________________ C. ______________________________

7. What is the best preparation for leisure in late life? (618)

____

8. What factors account for gains in well-being that result from leisure pursuits? (618)

____

____

9. With age, frequency and variety of leisure pursuits decline due to ________________ ________________. (618)

10. List characteristics associated with likelihood of volunteering. (619)

A. ______________________________ B. ______________________________

C. ______________________________ D. ______________________________

11. Older adults report (less/greater) awareness of public affairs and vote at a (lower/higher) rate than do other adults. (619)

Successful Aging

1. Recent definitions of successful aging focus not on specific ________________, but on the ________________ people use to reach personally valued goals. (619)

2. List nine ways in which older adults realize their goals while minimizing losses. (619-620)

A. ____

B. ____

C. ____

D. ____

E. ____

F. ____

G. ____

H. ____

I. ____

3. To better foster successful aging, reforms are necessary in public _________ that meet elders' needs, greater emphasis on life-long ________________, and preparation for greater numbers of ____________ elderly. (620)

ASK YOURSELF . . .

Compared to a few years earlier, 80-year-old Miriam took longer to get dressed in the morning. Joan, her paid home helper, said, "I'd like you to wait until I arrive before dressing. Then I can help you and it won't take so long." What impact is Joan's approach is likely to have on Miriam's personality? What alternative approach to helping Miriam would you recommend to Joan? (see text pages 599-600)

____
____
____

Many elders adapt effectively to negative life changes. List personal and environmental factors that facilitate this generally positive outcome. (see text page 602)

____
____
____

Although involvement in religion is fairly stable throughout adulthood, it takes on greater meaning and is linked to many positive outcomes in old age. Explain why, drawing on late-life physical and psychosocial development. (see text pages 596-599)

____
____
____

Sam lives by himself in the same home he has occupied for over 30 years. His adult children can't understand why he won't move across town to a modern apartment, which would be easier to care for than his old, dilapidated house. Cite as many possible reasons as you can think of that Sam prefers to stay where he is. (see text pages 606-608)

____
____
____

Vera, a nursing home resident, speaks to her adult children and a close friend on the phone every day. In contrast, she seldom attends scheduled social events or interacts with the woman who shares her semiprivate room. Using socioemotional selectivity theory, explain Vera's behavior. (see text pages 604-605)

____
____
____

Seventy-year-old Sean says his 40-year marriage to Caitlin is the happiest it's ever been. Caitlin agrees that she's more content than before, but she isn't quite as positive as Sean. What might account for Sean and Caitlin's high marital satisfaction and Sean's especially favorable response? (see text page 609)

____
____
____

Lottie, a never-married elder, lives by herself. Curt, a college student who just moved in next door, is certain Lottie must be very lonely. Why is Curt's assumption probably wrong? (see text page 612)

Mae, who lost her job at age 51 and couldn't afford her own apartment, moved in with her 78-year-old widowed mother, Beryl, who was glad to have Mae's companionship. Mae grew depressed and spent her days watching TV and drinking heavily. Although Beryl tried to be patient, she complained about Mae's failure to look for work. Mae became belligerent, pushed her mother against the wall, and slapped her. Explain why this mother–daughter relationship led to elder abuse. (see text pages 615-616)

Nate, happily married to Gladys, adjusted well to retirement. He also found that after he retired, his marriage became even happier. How can a good marriage ease the transition to retirement? How can retirement enhance marital satisfaction? (see text page 616)

SUGGESTED READINGS

Baltes, M. M. (1996). *The many faces of dependency in old age.* New York: Cambridge University Press. Experimental, observational, and intervention research is reviewed, shedding light on the development, maintenance, and reversibility of unnecessary dependency in old age. Shows how elders can use dependency strategically, in support of successful aging.

Decalmer, P., & Glendenning, F. (1997). *The mistreatment of elderly people* (2nd ed.). Thousand Oaks, CA: Sage. Presents a comprehensive overview of research on elder abuse. Perspectives from different disciplines on prevention, treatment, and legal issues are included.

McAdams, D. P., & de St. Aubin, E. (1998). *Generativity and adult development: How and why we care for the next generation.* Washington, DC: American Psychology Association. Explores the psychological, social, and cultural aspects of generativity – how adults construct their lives in ways that promote subsequent generations and work to leave an enduring legacy that will outlive the self.

Lopata, H. Z. (1996). *Current widowhood: Myths and realities.* Thousand Oaks, CA: Sage. Explores a range of issues related to widowhood, including emotions, identity, roles, and support systems. Also includes a historical and cross-cultural perspective.

PUZZLE 18.1 TERM REVIEW

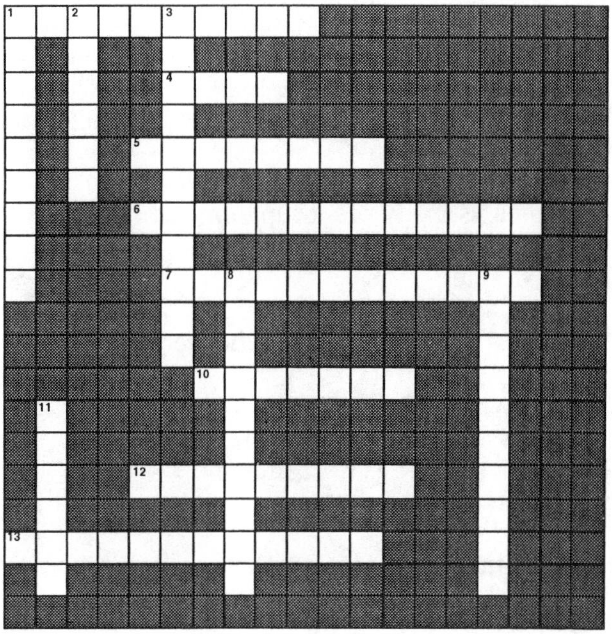

Across

1 Aging in which gains are maximized and losses minimized

4 _____ care communities: housing for the elderly offering a range of options, from independent or congregate housing to full nursing home care, guaranteeing that elders' needs will be met in one place as they age

5 Theory of aging which views the decline in social interaction in late adulthood as due to failure of the social environment to offer opportunities for social contact

6 Theory of aging which views the decline in social interaction in late adulthood as due to mutual withdrawal between elders and society in anticipation of death

7 _____-ignore script: typical pattern of interaction in which elders' independent behaviors are mostly ignored, leading them to occur less often

10 Negative outcome of Erikson's theory; feeling dissatisfied with one's life with too little time to make significant changes

12 Positive outcome of Erikson's theory; feeling whole, complete, and satisfied, having accepted one's life course

13 Process of telling stories about people and events from the past and reporting associated thoughts and feelings

Down

1 Friends who are not intimates but with whom the individual spends time occasionally

2 Social _____: Model of age-related changes in social networks, which views the individual within a cluster of relationships moving through life. Close ties are in the inner circle, less close ties on the outside. With age, people change places in the convoy, and some are lost entirely.

3 Socioemotional _____ theory: Theory of aging which views the decline in social interaction in late adulthood as due to physical and psychological changes leading elders to select associates on the basis of emotion. Consequently, they prefer familiar partners with whom they have developed pleasurable relationships.

8 _____-support script: typical pattern of interaction in which elders' dependency behaviors are attended to immediately, reinforcing those behaviors

9 Housing for the elderly that adds a variety of support services, including meals in a common dining room

11 Life _____: process of reflecting on the meaning of past experiences with the goal of increasing self-understanding

SELF-TEST

1. According to Erikson, elderly people (592-593)
   a. all achieve a sense of acceptance concerning death, known as integrity.
   b. all become bitter and hopeless when they realize their time is "up."
   c. feel bitter before they accept their mortality.
   d. will experience a psychological conflict between *a* and *b*.

2. Which is NOT one of the three distinct tasks involved in the conflict of ego integrity vs. despair, according to Peck? (593)
   a. ego differentiation vs. work-role preoccupation
   b. body transcendence vs. body preoccupation
   c. ego transcendence vs. ego preoccupation
   d. ego integrity vs. despair

3. According to Labouvie-Vief's view of psychosocial development, older adults (593-594)
   a. are more in touch with their feelings.
   b. react more strongly to their emotions.
   c. loose touch with their feelings over time.
   d. don't integrate objective and subjective aspects of emotion.

4. Which shift in personality characteristics during late adulthood is NOT mentioned in the text? (595)
   a. increase in agreeableness—generous, acquiescent, and good natured
   b. increase in honesty
   c. reduction in sociability
   d. increase in acceptance of change

5. Which is NOT true concerning older adults and religion? (596-599)
   a. over 76 percent of Americans 65 and older consider religion to be very important in their lives.
   b. over 50 percent attend religious services weekly
   c. men are more likely to be active members of churches than women
   d. nearly two-thirds watch religious TV programs

6. Older adults' dependency in Western society is (599-600)
   a. encouraged by the stereotype of the elderly as passive and incompetent.
   b. feared and considered by many older adults as a signal that it is time to die.
   c. often attributed to non-responsive social partners
   d. both a and b.

7. Depression in old age (601)
   a. is often lethal, because people age 65 and older have the highest suicide rate of any age group.
   b. is not very serious because it is so common among the elderly.
   c. rarely occurs, because elderly people are very well adjusted.
   d. is not likely to be noticed by family members, because elderly people are almost always depressed.

8. Extroverted elders are (602)
   a. less likely to take advantage of opportunities to socialize.
   b. more likely to be lonely
   c. more likely to be depressed.
   d. more likely to have high self-esteem and life satisfaction.

9. Which theory suggests that as individuals age, they gradually choose to reduce the number of close relationships that they are engaged in? (603-604)
   a. disengagement theory
   b. socioemotional selectivity theory
   c. activity theory
   d. withdrawal theory

10. Older adults report greater life satisfaction when they live (606)
   a. in urban areas close to hospitals and restaurants.
   b. in rural areas which are peaceful and close to nature.
   c. in neighborhoods where many senior citizens live.
   d. near family members, regardless of social support provided by nearby friends and neighbors.

11. Older adults are (606)
   a. more likely than other age groups to be the victims of violent crime.
   b. more likely to worry about becoming a victim of crime when they live alone in an inner-city area than they are to worry about health, income, and housing.
   c. less likely than other age groups to be the victims of pickpockets and purse snatchers.
   d. not more likely to worry about crime due to a few isolated incidents, because they are well-adjusted and realize how unlikely it is to happen.

12. The majority of older Americans want to live (606-607)
   a. in their own homes and neighborhoods, and almost 90 percent of them do.
   b. in their own homes and neighborhoods, but only 50 percent manage to.
   c. in a more temperate climate like that found in Florida and Arizona.
   d. with their children or other family members.

13. Which type of residential community offers elderly adults single-level independent living space with modifications such as grab bars in bathrooms, and support services such as meals available in a common dining room? (608)
   a. retirement communities
   b. congregate housing
   c. life care communities
   d. nursing homes

14. In the United States, how likely are marriages to last 50 years or more? (609)
   a. 1 in 100 (1 percent)
   b. 1 in 10 (10 percent)
   c. 1 in 5 (20 percent)
   d. 1 in 3 (33 percent)

15. Which is NOT true concerning divorce in late adulthood? (609-610)
    a. The divorce rate is rising among people age 65 and older.
    b. Divorce is harder on elderly women than elderly men.
    c. Elderly men typically mention lack of shared interests and activities as the reasons for their divorce.
    d. Elderly women typically mention a lack of sexual contact as the reason for their divorce.

16. Who is the most likely to remarry in late adulthood? (611)
    a. widowed men
    b. widowed women
    c. divorced men
    d. divorced women

17. Elderly adults who never marry and never have children (612)
    a. are more likely than other elders to have siblings, other relatives, and nonrelatives living in their household.
    b. are likely to develop alternative meaningful relationships.
    c. are likely to be alone and forsaken.
    d. both a and b

18. Most elderly siblings (612)
    a. live within 30 miles of each other.
    b. consider bonds with their brothers to be closer than bonds with sisters.
    c. rarely talk to each other.
    d. visit each other at least several times a year.

19. Children, grandchildren, and great-grandchildren provide older adults (614-615)
    a. with a wider potential network of support.
    b. with a very gratifying link between themselves and the future.
    c. with relationships that are often closer on the paternal side.
    d. with both a and b.

20. Which is the most common form of elder maltreatment suggested by the number of reported cases? (615)
    a. physical neglect
    b. financial abuse
    c. psychological abuse
    d. physical abuse

21. To help prevent elder maltreatment (616)
    a. negative stereotypes of aging should be countered.
    b. legal action is necessary in all cases.
    c. prevention programs offering counseling and education for caregivers and relief services should be developed and used.
    d. both a and c.

22. American retirees (616-617)
    a. re-enter the labor force within 1 year of initial retirement one-third of the time.
    b. are waiting longer to initially retire.
    c. must retire by the mandatory retirement age of 75.
    d. both a and b.

23. Which is NOT a factor associated with reluctance to retire? (617)
    a. high work involvement
    b. financial worries
    c. having a high-stress job
    d. loss of social contacts

24. Involvement in leisure activities (618-619)
    a. is related to elders' physical, but not psychological, well-being.
    b. does not change much over the course of a lifetime.
    c. increases in frequency and variety with age.
    d. become less sedentary and less home based after age 75.

25. Which of the following does not characterize a successful ager? (619)
    a. effectively copes with physical changes and negative life events
    b. promotes self-acceptance and pursuit of hoped-for possible selves
    c. has high-quality relationships, which offer social support and pleasurable companionship
    d. reacts with despondency to aging and other losses

CHAPTER 19

DEATH, DYING, AND BEREAVEMENT

BRIEF CHAPTER SUMMARY

Although people generally wish for a quick, peaceful death, it is relatively rare. Death is usually is a long, drawn out process. While brain death is the accepted definition of the end of life in industrialized nations, controversy continues when patients are incapable of conscious functioning or when suffering patients ask to die. People can best insure that their wishes concerning medical treatment will be honored by communicating acceptable treatments or appointing a person responsible for medical decisions through an advanced medical directive.

Our understanding of death progresses gradually in childhood. While the death concept is grasped by adolescence, it is not yet fully applied to everyday reality. Both children and adolescents benefit from open, honest communication about death, and death anxiety declines with age. Kübler-Ross's stage theory provided structure to our understanding of the psychological side of dying. However, many factors contribute to the experience of dying, which varies considerably. Although most people want to die at home, it is rarely feasible, and the pros and cons must be carefully weighed. Yet dying in a hospital can be depersonalizing. Hospice care, with its focus on the needs of the patient and family, can make death more comfortable and consistent with the context of the patient's life.

Grieving usually takes place in three stages in which the reality of the death is avoided, the loss is confronted and felt intensely, and, finally, attention is shifted toward fostering new goals and relationships. Personal, cultural, and situational factors, such as how suddenly and at what stage of life the death occurs, influence the grieving process. The death of a child is especially difficult for parents to weather, and grieving is also extended for children who lose a parent or sibling. Bereaved individuals require empathy and understanding and can benefit from attending self-help groups or grief therapy. Death education is helpful when it helps people confront their own mortality. Being in touch with death, although sometimes disturbing, can help us appreciate life and live more fully.

LEARNING OBJECTIVES

After reading this chapter, you should be able to:

19.1. Describe the physical changes of dying, along with their implications for defining death and for the meaning of death with dignity. (628-630)

19.2. Discuss age-related changes in conceptions of and attitudes toward death, and cite factors that influence death understanding and anxiety. (630-634)

19.3. Describe and evaluate Kübler-Ross's stage theory, citing contextual factors that influence the responses of dying patients. (634-638)

19.4. Evaluate the extent to which homes, hospitals, and the hospice approach meet the needs of dying people and their families. (638-640)

19.5. Discuss ethical and legal controversies surrounding euthanasia and assisted suicide. (640-645)

19.6. Describe the phases of grieving, factors that underlie individual variations, and bereavement interventions. (645-649)

19.7. Explain how death education can help people cope with death more effectively. (649-651)

STUDY QUESTIONS

How We Die

1. Of the ______________ of people in industrialized nations who die suddenly (within a few hours of symptoms) ____________________ percent are victims of ________________________. (628)

2. Death is long and drawn out for _______________ of people, many (more/less) than in the past. (628)

3. Match each of the following stages of death with the appropriate description. (629)

____	Gasps and muscle spasms during the moments in which the body first cannot sustain life.	1. Mortality
____	A short interval in which heart beat, circulation, breathing, and brain functioning stop, but resuscitation is still possible.	2. Agonal phase
____	Permanent death. Within a few hours, the body appears shrunken, no longer like the person was when alive.	3. Clinical death

4. Death is a(n) (event/process). What is the current definition of death used in most industrialized nations? (629)

____

5. Explain why no legal definition of death exists in Japan, leaving doctors to rely on the absence of heartbeat and respiration to define it. (629)

____

6. Because approximately 10,000 people are in a ________________________ state, many experts believe that the absence of activity in the __________________ should be sufficient to declare death. (629)

7. Most people cannot be granted an easy end. However, they can die with dignity and integrity. Briefly list three ways in which this can be fostered. (629-630)

A. ____
B. ____
C. ____

Understanding of and Attitudes Toward Death

1. List the three ideas on which a realistic understanding of death is based. (631)
A. ______________________________ B. ______________________________
C. ______________________________

2. Without clear explanations, preschoolers rely on __________________ and ________________ thinking to make sense of death and can arrive at incorrect conclusions about what causes death. (631)

3. Most children grasp the three components of the death concept by age ____. (631)

4. Children's ____________________ with death and ethnic variations in ____________________ teachings can influence children's understanding of death. (631)

5. True or False: Discussing death honestly and directly is likely to increase children's fears. (631)

6. Although adolescents grasp the permanence and nonfunctionality of death, they often describe it as an enduring ________________ state and formulate personal theories of ________________________________. (632)

7. Adolescents' understanding that death can occur any time does not make them more ___________-conscious, evidence that they do not apply their knowledge of death to their own lives. (632)

8. List three reasons why adolescents have difficulty integrating logic with reality in the domain of death.
A. ______________________________
B. ______________________________
C. ______________________________

9. List five ways to discuss death with adolescents. (633)
A. ______________________________
B. ______________________________
C. ______________________________
D. ______________________________
E. ______________________________

10. In early adulthood, many people ________________________ thoughts of death. Middle-aged people no longer have a ____________ conception of their own death. In late adulthood, adults think and talk (more/less) about death. Compared to middle-aged adults, elders' thoughts are more focused on the __________________ of death than on the ______________ of death. (633)

11. Research reveals (few/large) individual and cultural differences in the aspects of death which cause distress. (634)

12. Identify two personal factors which limit death anxiety. (634)
A. ______________________________
B. ______________________________

13. Death anxiety (increases/declines) with age. (634)

14. In a study of Israeli adults, modes of ______________________________, the sense that one will continue to live through one's children or through one's work or personal influence, predicted (reduced/increased) fear of death. This was especially true among adults with __________________ attachments. (634)

15. (Women/Men) are more anxious about death. (634)

16. People who are _________________ or generally _________________ are more likely to have greater death concerns. A large gap between their __________________ and ______________ selves leaves them with a sense of incompleteness when they contemplate death. (634)

17. True or False: Children rarely display death anxiety. (634)

Thinking and Emotions of Dying People

1. List Kübler-Ross's five stages of dying, along with a brief suggestion for how family members and health professionals should react at each of the first four stages. (635-636)
A. ____
B. ____
C. ____
D. ____
E. ____

2. True or False: Strict interpretation of Kübler-Ross's stages has helped professionals provide more sensitive care to dying patients. (636)

3. Rather than stages, the five reactions Kübler-Ross observed are better viewed as ______________________________________ that anyone may call on in the face of stress. Furthermore, her list is ____________________________. (636)

4. True or False: Kübler-Ross's theory views dying patients' reactions in relation to the contexts that give them meaning. (636)

5. An *appropriate death* is one that makes sense in terms of the individual's ____________________________________ and _____________, preserves or restores important _________________________, and is as free of suffering as possible. (636)

6. True or False: The nature of the disease has little to do with the patient's response to dying. (636)

7. Views of stressful life events and coping styles applied in the past (are/are not) related to individuals' approach to and view of dying. (636)

8. Why do some people pretend that a dying person's condition is not as bad as it is? (637)

____

9. List three keys to success in training health care professionals to respond to the psychological needs of dying patients and their families. (637)
A. ____
B. ____
C. ____

10. Dying patients often move through a hope trajectory—first hoping for a ____________________, then for ____________________________, and finally for a ______________________________. (637)

11. List four suggestions for how to communicate effectively with the dying. (637)
A. ______________________________
B. ______________________________
C. ______________________________
D. ______________________________

12. A strong sense of ____________________ reduces fear of death. Cultural variations include Buddhism, which emphasizes that death leads to ____________________, the common Native-American view of life and death as ____________________, and the African-American approach in which the dying of a loved one ________________ family members. (637-638)

A Place to Die

1. List three or more reasons why it is difficult to honor terminally ill patients wishes to die at home. (638)

____
____
____

2. What type of deaths typically occur in hospital emergency rooms? What is needed to help survivors cope with this type of death? (639)
A. ______________________________
B. ______________________________

3. Dying on an intensive care ward can be very depersonalizing because ____________________ and communicating with ____________________ are second to the monitoring of the patient's condition. (639)

4. ______________ patients account for most cases of prolonged dying, and typically die in general or specialized ____________________ hospital units. (639)

5. Central to the hospice approach is that the dying person and his or her family should be offered ______________ that guarantee an ______________ death. Provide three examples of hospice approaches. (639)
A. ____________________ B. ____________________
C. ____________________

6. Hospice care (is/is not) affordable for most dying patients. At present, the majority of Americans (are/are not) familiar with the hospice philosophy. (639)

The Right to Die

1. True or False: In the United States, a uniform right-to-die policy exists for cases of terminal illness and persistent vegetative state. (640)

2. When there is no hope of recovery, ______ percent of Americans support the patients' right and ______ percent support the family members' right to passive euthanasia. (640)

3. How can people best ensure that their wishes will be followed should they become incurably ill? (641)

____

4. A __________________ specifies the treatments a person does or does not want in the event of a near-death situation, and may include a request for ______________ care in the event of severe pain. (641)

5. Living wills rarely apply to conditions which are not classified as terminal. List two such conditions. (642)

A. ____

B. ____

6. The ____ is more flexible than the living will because it is not limited to terminal illnesses. (642)

7. Less than _____ percent of Americans have executed advanced medical directives such as those addressed above. In a few states, those who do not do so can be covered through the appointment of a ___________________________. (642)

8. About _____ percent of polled Dutch, _____ percent of American, and _____ percent of Canadian citizens supported voluntary active euthanasia under certain conditions. (642)

9. Supporters of voluntary euthanasia believe that it represents the most _________________________________ option in the case of severely painful terminal illness. Opponents stress the difference between ________________ and ________________, argue that it will undermine people's ______________ in doctors, and that it could lead to a ____________________ of acceptable euthanasia. (642-643)

10. In the Northern Territory of Australia, a law allowing voluntary active euthanasia was (overturned/upheld) by the Australian Parliament after being (well accepted/heavily criticized). (644)

11. In the Netherlands, exemptions from punishment for doctors who engage in voluntary active euthanasia lead to the unrequested death of ___________ patients in a single year. (644)

12. (Less/More) public support exists for assisted suicide than for euthanasia; about _____ percent of Americans approve. However, in a study of 60- to 100-year olds, only ___________ said they would end their lives. Those who indicated they would choose assisted suicide tended to be ________________ and more often __________________________ than __________________________. (643)

13. Under what four conditions did a panel of medical experts conclude that assisted suicide is warranted? (645)

A. ____

____

B. ____

____

C. ____

____

D. ____

____

Bereavement: Coping with Death of a Loved One

1. ____________________ is the experience of losing a loved one by death. ____________________ is the culturally specified expression of the associated thoughts and feelings. (645)

2. Match each of the following phases of grieving with the appropriate descriptions. (645-646)

____	Attention shifts to the surrounding world	1. Avoidance
____	Grief is experienced most intensely	2. Confrontation
____	Shock followed by disbelief, lasting from hours to weeks	3. Accommodation
____	Relationship with deceased is transformed from physical presence to inner representation	
____	A numbed feeling serves as "emotional anesthesia"	
____	Mourner begins to face the reality of the loss	

3. List five factors which influence the duration of and reactions within each phase of grieving. (646)

A. ______________________________ B. ______________________________
C. ______________________________ D. ______________________________
E. ______________________________

4. ____________________ is usually pronounced after sudden death, while ____________________ grieving occurs in the case of prolonged death. Understanding the ____________________ makes adjustment easier for survivors. (646)

5. People grieving a suicidal loss are (more/less) less to blame themselves for what happened and recovery is quite (brief/prolonged). (646)

6. List three reasons why the death of a child is the most difficult loss an adult can face. (646-647)

A. ______________________________
B. ______________________________
C. ______________________________

7. Parents who have lost a child (often/rarely) report considerable distress many years later. When the marital relationship was previously unsatisfactory, the death of a child is more likely to lead to ________________. (647)

8. Children describe frequent crying, difficulty concentrating and sleeping, headaches, and other physical symptoms several ____________ to __________ after a family death. (Many/Few) said they actively maintained mental ____________ with their dead parent or sibling. (647)

9. Young children need careful, repeated explanations assuring them that the deceased parent did not ______________________________ and was not ______________. (647)

10. Explain why teenagers are more likely to become depressed or to escape from grieving through acting-out behavior. (647)

____

11. Younger individuals display (more/less) negative outcomes than most widowed elders. Briefly explain why this is true. (648)

____

____

12. Identify three groups at increased risk for bereavement overload, and circle the letter of the one most prepared to handle it. (648)

A. ____________________ B. ____________________

C. ____________________

13. Match the following religious or cultural groups with the appropriate descriptions of mourning behavior (some apply to more than one group). (650-651)

_____ Extensive ritual accompanies funeral and burial	1. Jews
_____ Among the least ritualized of ceremonies	2. Christians
_____ Venting of deep emotions is expected	3. Quaker
_____ Usually restrained in show of emotion	4. African-American
_____ Actively discourage show of any emotion	5. Balinese of Indonesia
_____ Affirm personal survival without specifying form	6. Tribal and village cultures
_____ Focus on here and now "salvation by character" rather than after life	
_____ Include elaborate beliefs about ancestor spirits	

14. Identify four ways in which funerals and memorial services assist the bereaved. (651)

A. ____________________ B. ____________________

C. ____________________ D. ____________________

15. Mourners report that they (do/do not) get enough support. Identify contacts that bereaved people are likely to view positively and those they view negatively. (648)

Positive: ______________________________

Negative: ______________________________

16. List five suggestions for resolving grief after a loved one dies. (649)

A. ______________________________

B. ______________________________

C. ______________________________

D. ______________________________

E. ______________________________

17. True or False: Research supports the effectiveness of self-help groups in reducing stress, reworking marital ties after the death of a child, and providing peer support to teens. (649)

18. Under what conditions is grief therapy more likely to be necessary? (646)

A. ______________________________

B. ______________________________

C. ______________________________

C. ______________________________

Death Education

1. Using a ______________ approach to death education increases knowledge, but often leaves students more uncomfortable about death. ________________ programs that help people confront their own mortality are less likely to increase death anxiety. (650)

2. Being in touch with death and dying allows us to live ________________. (651)

ASK YOURSELF . . .

When 4-year-old Chloe's aunt died, Chloe asked, "Where's Aunt Susie?" Her mother explained, "Aunt Susie is taking a long, peaceful sleep." For the next 2 weeks, Chloe refused to go to bed, and when finally coaxed into her room, she lay awake for hours. Explain the likely reason for Chloe's behavior, and suggest a better way of answering Chloe's question. (see text page 630-632)

____

____

____

Explain why older adults think and talk more about death than do younger people but nevertheless feel less anxious about it. (see text page 634)

____

____

____

Reread the description of Sofie's mental and emotional reactions to dying on page 635. Then review the story of Sofie's life on page 4. To what extent did Sofie follow Kübler-Ross's stages? How were Sofie's responses consistent with her personality and lifelong style of coping with adversity? (see text pages 635-636)

____

____

____

When 5-year-old Timmy's kidney failure was diagnosed as terminal, his parents could not accept the tragic news. Their hospital visits became shorter, and they evaded his anxious questions. Timmy tried to figure out what he had done to drive his parents away. Eventually, he blamed himself. He died with little physical pain but very much alone, and his parents suffered prolonged guilt after his death. Explain how hospice care could have helped Timmy and his family. (see text page 649)

____

____

____

Noreen knows that the majority of people die gradually. Thinking ahead to the day she dies, she imagines a peaceful scene in which she says good-bye to loved ones. What social and medical practices are likely to increase Noreen's chances of dying in the manner she desires? (see text pages 629-630)

____

____

____

If he should ever fall terminally ill, Ramon is certain that he wants doctors to halt life-saving treatment. To best ensure that his wish will be granted, what should Ramon do? Why is it impossible to guarantee that Ramon's desires will be honored in the United States at this time? (see text pages 641-642)

Compare the phases of grieving with terminally ill patients' thoughts and feelings as they move closer to death, described on pages 635-636. Can a dying person's reactions be viewed as a form of grieving? Explain. (see text pages 645-646)

Cite examples of age-graded, history graded, and nonnormative influences on the grieving process, referring to research findings on pages 646-648. If you need to review these contexts for lifespan development, return to Chapter 1, pages 11-12.

Age-graded:

History graded:

Nonnormative:

Explain how death education can help people cope with death more effectively. (649-651)

SUGGESTED READINGS

Kastenbaum, R. (1998). *Death, society, and human experience* (6th ed.). Needham Heights, MA: Allyn & Bacon. Provides an interdisciplinary approach to understanding death and the dying experience, through the use of case examples and reflective exercises. A landmark text in death education.

Mannino, J. D. (1997). *Grieving days, healing days*. Needham Heights, MA: Allyn & Bacon. Humanizes the experiences of death and grieving, and promotes a more healthy, well-informed response to the many ways death enters our lives. This manual contains over 130 exercises and activities that explore the emotional, social, and legal aspects of death.

Pojman, L. P. (2000). *Life and death: Grappling with the moral dilemmas of our time* (2nd ed.). Belmont, CA: Wadsworth. Explores a range of moral issues surrounding matters of life and death, including suicide, euthanasia, abortion, and the death penalty.

Scherer, J. M., & Simon, R. J. (1999). *Euthanasia and the right to die: A comparative view*. Lanham, MD: Roman & Littlefield. Examines euthanasia and the right to die in terms of the social, legal, and religious contexts in various countries. Discusses the general public's approval of physician-assisted suicide versus the illegality of the practice.

PUZZLE 19.1 TERM REVIEW

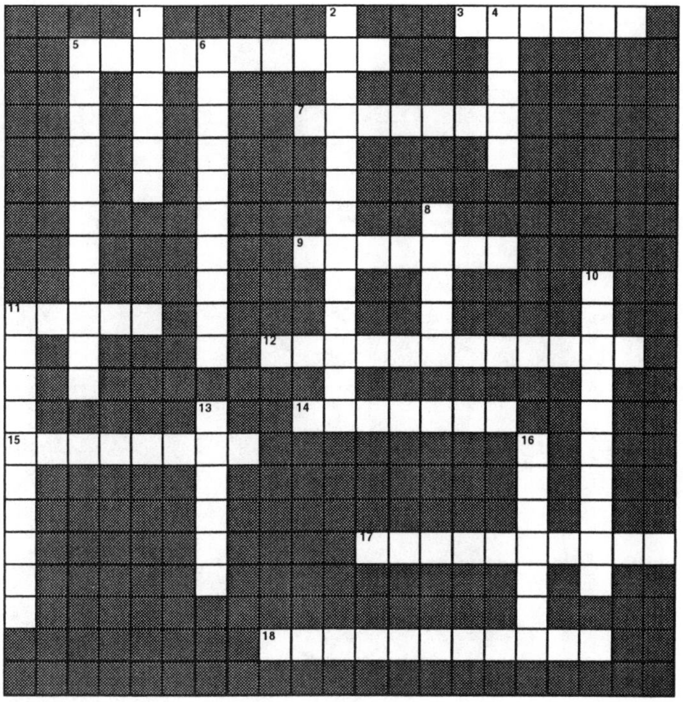

Across

3 Phase of dying in which gasps and muscle spasms occur during the moments in which the body first can no longer sustain life

5 Kübler-Ross's stage of dying, in which the person, in the last weeks or days, enters a state of peace and quiet about upcoming death

6 Comprehensive program of support services that focuses on meeting terminally ill patients' physical, emotional, social, and spiritual needs and offers follow-up services to families

9 _____ medical directive: written statement of desired medical treatment should one become incurably ill

10 _____ death: irreversible cessation of all activity in the brain and brain stem; definition of death accepted in most industrialized nations

12 Component of the death concept specifying that all living things eventually die

14 Death _____: fear of death

15 Durable power of _____ for health care: written statement authorizing appointment of another person (usually a family member) to make health care decisions on one's behalf in case of incompetence

17 Kübler-Ross's stage of dying, in which the person becomes depressed, grieving the ultimate loss of his or her life

18 Experience of losing a loved one by death

Down

1 Voluntary _____ euthanasia: practice of ending a patient's suffering, at their request, before a natural end to life; a form of mercy killing

2 Third phase of grieving, in which it subsides and emotional energies are freed for meeting everyday responsibilities, investing in new activities, and seeking the companionship of others

4 Intense physical and psychological distress following a loss

5 Death that makes sense in terms of pattern of living and values, preserves or restores significant relationships, and is as free of suffering as possible

6 Component of the death concept specifying that once a living thing dies, it cannot be brought back to life

8 Kübler-Ross's stage of dying, in which the person expresses anger at having to die without a chance to do all he or she wants to do

10 Practice of ending the life of a person suffering from an incurable condition

11 Kübler-Ross's stage of dying, in which the person attempts to forestall death by bargaining for extra time—a deal he or she may try to strike with family members, friends, caregivers, or God

13 Kübler-Ross's stage of dying, in which the person denies the seriousness of his or her terminal condition in an effort to escape from the prospect of death

16 _____ euthanasia: practice of withholding or withdrawing life-sustaining treatment, permitting a patient to die naturally

PUZZLE 19.2 TERM REVIEW

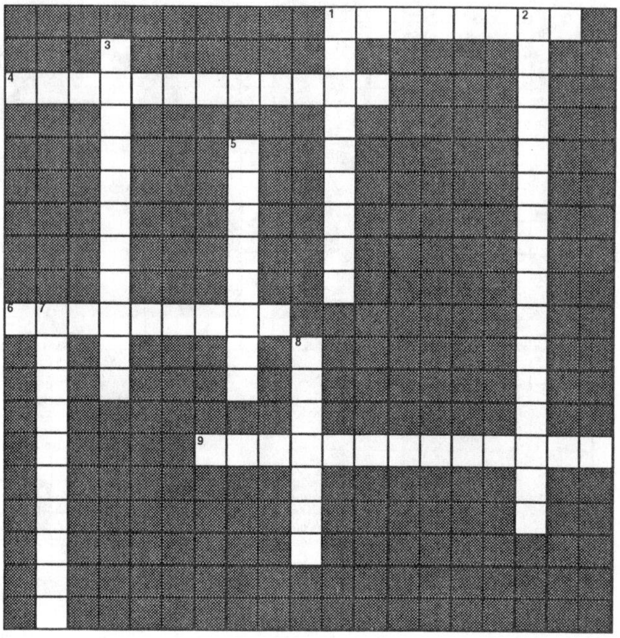

Across

1 Culturally specified expression of bereaved people's thoughts and feelings through funerals and other rituals

4 _____ grieving: before a prolonged, expected death acknowledging that the loss is inevitable preparing emotionally for it

6 First phase of grieving, in which the survivor experiences shock followed by disbelief

9 Second phase of grieving, in which the reality of the loss is confronted and grief is experienced most intensely

Down

1 Phase of dying in which the individual passes into permanent death

2 Component of the death concept specifying that all living functions, including thought, feeling, movement, and body process, cease at death.

3 ____ ____: written statement specifying treatments one does or does not want in case of a terminal illness, coma, or other near-death situation (2 words)

5 ____ death; phase of dying in which heart beat, circulation, breathing, and brain functioning stop, but resuscitation is still possible

7 Persistent _____ state: produced by absence of brain wave activity in the cortex; the person is unconscious, displays no voluntary movements, and has no hope of recovery

8 _____ care: provides sufficient, often potent medication to relieve pain for terminally ill, suffering patients even though the treatment might shorten life

SELF-TEST

1. Most people in industrialized nations die (628)
   a. in hospitals.
   b. with loved ones attending their last moments.
   c. alone in their homes.
   d. in public venues.

2. _____percent of the people in industrialized nations die suddenly. The rest suffer long and drawn out deaths. (628)
   a. 5
   b. 10
   c. 15
   d. 25

3. The agonal phase of dying (629)
   a. is the first of the three stages generally associated with dying.
   b. is derived from the term agony which means extreme pain.
   c. refers to gasps and muscle spasms during the moments in which the body first cannot sustain life anymore.
   d. is both a and c.

4. Which is used to determine the end of life (death) in most industrialized nations? (629)
   a. brain death
   b. persistent vegetative state
   c. absence of heartbeat and respiration
   d. absence of activity in the brain stem

5. Death with dignity does NOT refer to (629-630)
   a. a quick, agony-free death during sleep.
   b. personal control over this final phase of life.
   c. knowing the truth about the disease or diagnosis.
   d. an unexpected death.

6. Compared to earlier generations, children and adolescents in today's developed nations are, for the most part (630)
   a. insulated from death.
   b. less uneasy about death.
   c. more aware of death due to television and movie exposure.
   d. less in denial about death.

7. Which of the following is NOT one of the three ideas on which a realistic death is concept based? (631)
   a. permanence
   b. universality
   c. totality
   d. nonfunctionality

8. Why doesn't teenage behavior reflect their understanding of death? (632)
   a. The adolescent personal fable leads them to believe that they are beyond death when engaged in risky behavior.
   b. The dividing line between life and death for adolescence is very sharp.
   c. Their bodies are growing in size and strength which is the opposite of death.
   d. Both a and c are true.

9. Which is NOT true concerning death anxiety in Western cultures? (634)
   a. Women are more anxious about death than men.
   b. Devout Christians are less fearful of death than devout atheists.
   c. The prevalence of death anxiety declines as people age.
   d. People with mental health problems are more likely to fear death.

10. Which is NOT one of the five reactions which characterize dying people, according to Elisabeth Kübler-Ross? (635-636)
    a. denial
    b. betrayal
    c. bargaining
    d. acceptance

11. Kübler-Ross's stages (636)
    a. are sequential steps that a "normal" dying person follows.
    b. almost always happen in the order she suggested.
    c. are actually five reactions that should be considered coping strategies.
    d. are reactions that characterize dying people.

12. Which of the following factors is NOT listed in the text as a factor affecting dying individuals' reactions to their impending death? (636)
    a. personality and coping style
    b. nature of the disease
    c. family members' and health professionals' behavior
    d. gender of the individual

13. Buddhists believe that (637)
    a. death is met with stoic self-control.
    b. a dying loved one signals a crisis that unites family members in caregiving.
    c. a dying person should be left alone.
    d. it is possible to reach Nirvana, a state beyond the world of suffering.

14. Which culture emphasizes the circular relationship between life and death as well as the importance of making way for others, and teaches its members to meet death with stoic self-control? (637-638)
    a. Western European
    b. Eastern European
    c. Native American
    d. African American

15. Nine out of ten Americans would prefer to die (638)
    a. in a hospital.
    b. in a hospice.
    c. at home.
    d. alone.

16. The central belief of the hospice approach is (639)
    a. that choices should be made for the dying person and his or her family.
    b. that quality of life is the most important issue surrounding one's journey toward death.
    c. people should always be in a hospital when they die.
    d. to prolong life.

17. Many of the people who are dying from _______ are served by hospice programs. (639)
    a. heart attacks
    b. cancer
    c. AIDS
    d. both b and c

18. Euthanasia refers to (640)
    a. the rights of young people to control society.
    b. the practice of ending the life of a person who is incurably ill.
    c. the removal of vital organs for transplants from brain dead individuals.
    d. the decision to die.

19. Which of the following is NOT a statement of desired medical treatment in the event of incurable illness? (641-642)
    a. an advance medical directive
    b. a living will
    c. durable power of attorney for health care
    d. a medical waiver

20. Which is NOT a condition that must be met before assisted suicide is warranted, according to one group of medical experts as cited in the text. (645)
    a. The patient requests assisted suicide repeatedly and freely and is suffering intolerably, with no satisfactory options.
    b. The doctor thoroughly explores alternatives for comfort care with the patient.
    c. When the doctor or patient believes that there are no other acceptable choices.
    d. The practice is consistent with the doctor's fundamental values.

21. The intense physical and psychological distress associated with the loss of a loved one is known as (645)
    a. mourning
    b. grief
    c. bereavement
    d. depression

22. Which is NOT one of the three phases of grieving? (645-646)
    a. avoidance
    b. anger
    c. confrontation
    e. accommodation

23. Adjusting to death is easier when (646)
    a. the survivor understands the reasons for it.
    a. the death was sudden.
    b. the death was a suicide.
    d. a child dies.

24. Among the best ways a nongrieving individual can help the bereaved is (648)
    a. by listening patiently and "just being there."
    b. by encouraging recovery.
    c. by giving advice based on one's own experience.
    d. encourage the bereaved to be alone.

25. According to the text, which is NOT a goal of death education? (650)
    a. preparing students to be informed consumers of medical and funeral services.
    b. promoting understanding of social and ethical issues involving death.
    c. increasing students' understanding of the physical and psychological changes that accompany dying.
    d. teaching students how to avoid grief.

ANSWERS TO PUZZLES

Puzzle 1.1

P					I									C	O	H	O	R	T	
S	E	Q	U	E	N	T	I	A	L			P		O					H	
Y					T							R		N		C			E	
C		M			E	X	P	E	R	I	M	E	N	T	A	L			O	
H		E			R							F		I		I			R	
O	B	S	E	R	V	A	T	I	O	N		O		N		N			Y	
A		O			I					A		R		U		I				
N		S			E					T		M		O		C	A	S	E	
A		Y			W			E		U		A		U		A				
L		S						T		R		T		S		L		R		
Y		T						H		E		I						E		
T		E						O			S	O	C	I	O			S		
I		M			N	O	B	L	E			N						I		
C								O				I						L		L
								G				S	E	N	S	I	T	I	V	E
	C	H	R	O	N	O	S	Y	S	T	E	M						E		A
											X					C		N		R
			M	A	T	U	R	A	T	I	O	N				R		C		N
								G								O		Y		I
N	O	R	M	A	T	I	V	E								S				N
											P	R	O	C	E	S	S	I	N	G

Puzzle 1.2

L	O	N	G	I	T	U	D	I	N	A	L								R	
I		O							A						S	O	C	I	A	L
F		N				C	O	N	T	E	X	T	S		E				N	
E		N							U						X				D	
S		O		H	I	S	T	O	R	Y		T	A	B	U	L	A		O	
P		R							A						A				M	
A		M		C		E	C	O	L	O	G	I	C	A	L					
N		A		O					I											
	S	T	A	G	E				S	T	A	B	I	L	I	T	Y			
C		I		N			D		T											
O		V		I			E		I	N	D	E	P	E	N	D	E	N	T	
E		E		T			P		C								T			
F				I			E				M	I	C	R	O		H			
F				V			N										N			
I				E			D	I	S	C	O	N	T	I	N	U	O	U	S	
C							E										G			
I			H	U	M	A	N			B	E	H	A	V	I	O	R	I	S	M
E							T										A			A
N	U	R	T	U	R	E			A	P	P	L	I	E	D		P			C
T																	H			R
		C	O	R	R	E	L	A	T	I	O	N	A	L			Y			O

Puzzle 2.1

					P					G								C		
					U				H	E	T	E	R	O	Z	Y	G	O	U	S
					B					N								D		
					L					E		R	E	A	C	T	I	O	N	
					I				M									M		
		S	U	B	C	U	L	T	U	R	E							I		
									T			P	O	L	Y	G	E	N	I	C
		I							A									A		
		D	O	M	I	N	A	N	T								D	N	A	
		E							I									C		
G	E	N	O	T	Y	P	E		O						A			E		
A		T				H			N						U					
M		I				E						F	R	A	T	E	R	N	A	L
E		C				N									O		E			
T		A				O			M	E	I	O	S	I	S		C			
E		L				T			I						O		E			
S					Z	Y	G	O	T	E					M		S			
						P			O						E		S			
						E			S				S	E	S		I			
									I								V			
H	O	M	O	Z	Y	G	O	U	S			C	A	R	R	I	E	R		

Puzzle 2.2

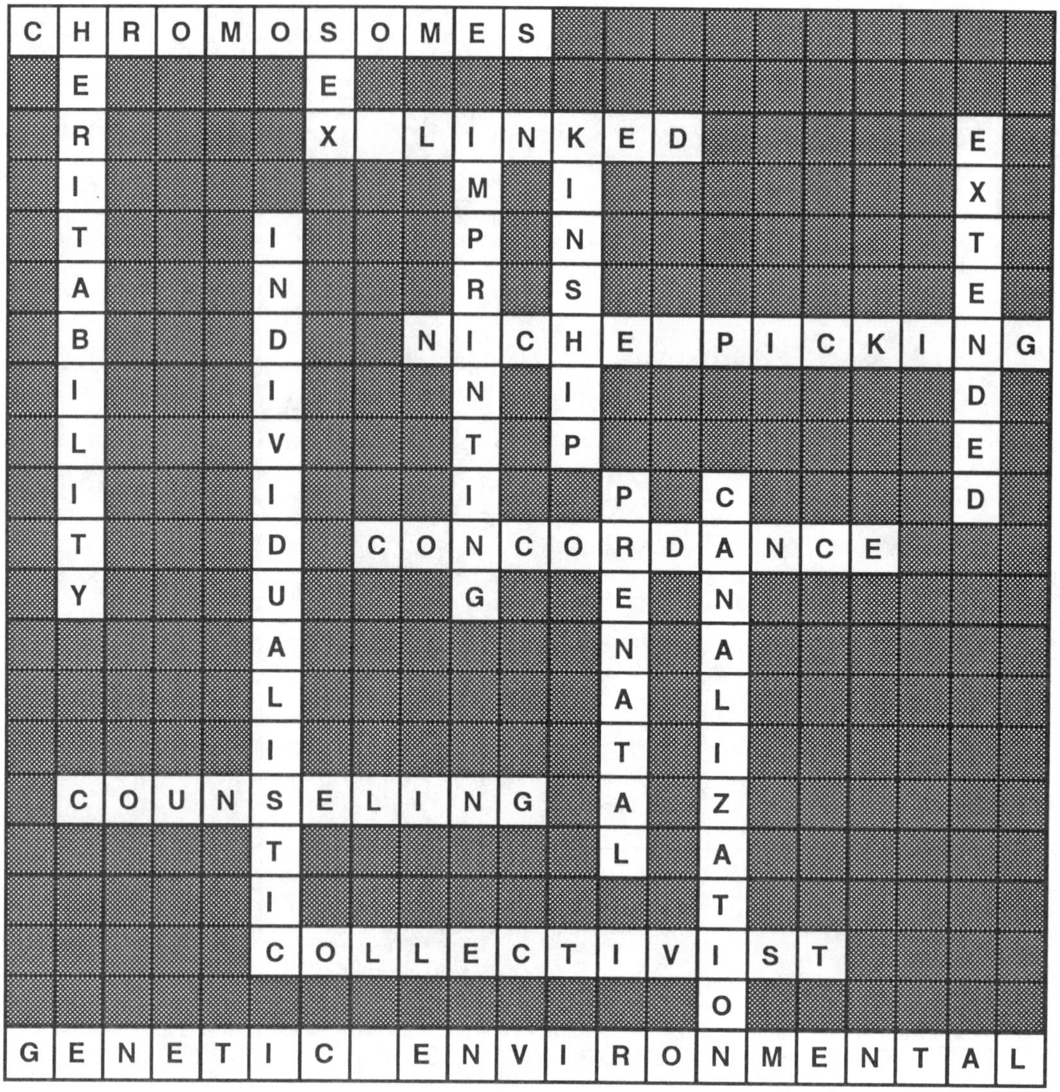

CHROMOSOMES
HERITABILITY
SEX LINKED
IMPRINTING
KINSHIP
EXTENDED
INDIVIDUALISTIC
NICHE PICKING
PRENATAL
CANALIZATION
CONCORDANCE
COUNSELING
COLLECTIVIST
GENETIC ENVIRONMENTAL

Puzzle 3.1

					V			C												
	T	R	I	M	E	S	T	E	R	S				R	E	F	L	E	X	
	E				R			S										M		
	R				N			A										B		
	A				I			R	E	M								R		
	T				X			E					A	C	U	I	T	Y		
	O							A			I			H				O		
	G							N	R	E	M			O						
	E										P		P	R	E	T	E	R	M	
	N	A	T	U	R	A	L				L			I						
				M							A			O						
				B					A	M	N	I	O	N						
A	N	O	X	I	A						T									
				L					P	L	A	C	E	N	T	A				
				I							T									
	D			C							I									
	A	P	G	A	R					M	O	R	T	A	L	I	T	Y		
	T			L							N									
	E																			

Puzzle 3.2

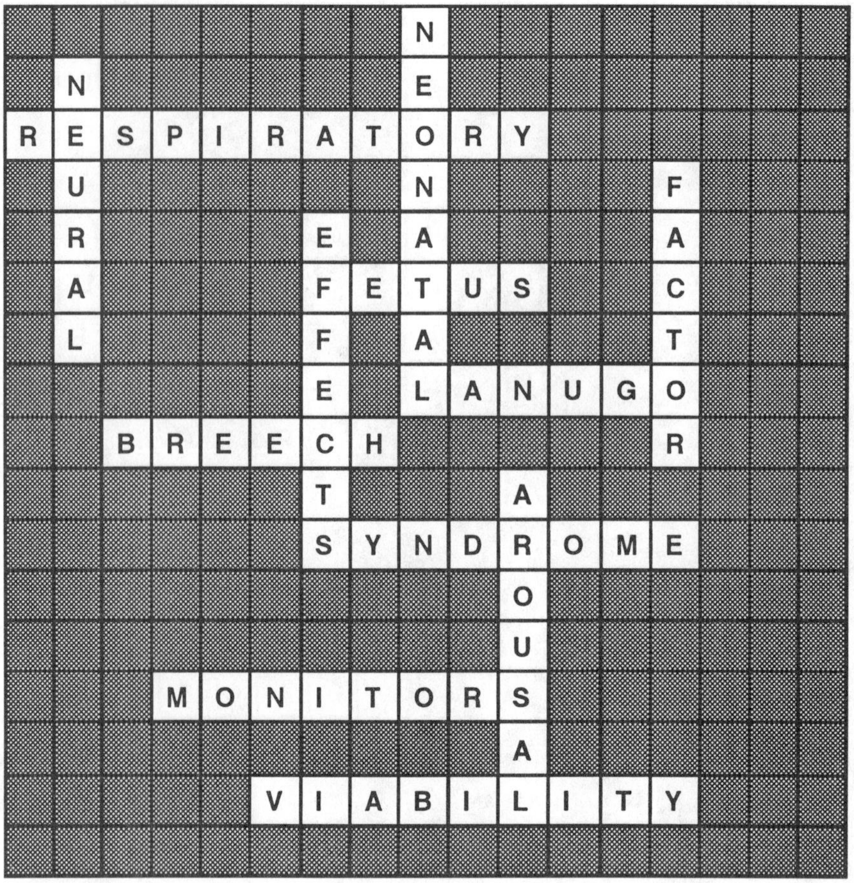
RESPIRATORY
FETUS
LANUGO
BREECH
SYNDROME
MONITORS
VIABILITY
NEURAL
NEONATAL
EFFECTS
FACTOR
AROUSAL

Puzzle 4.1

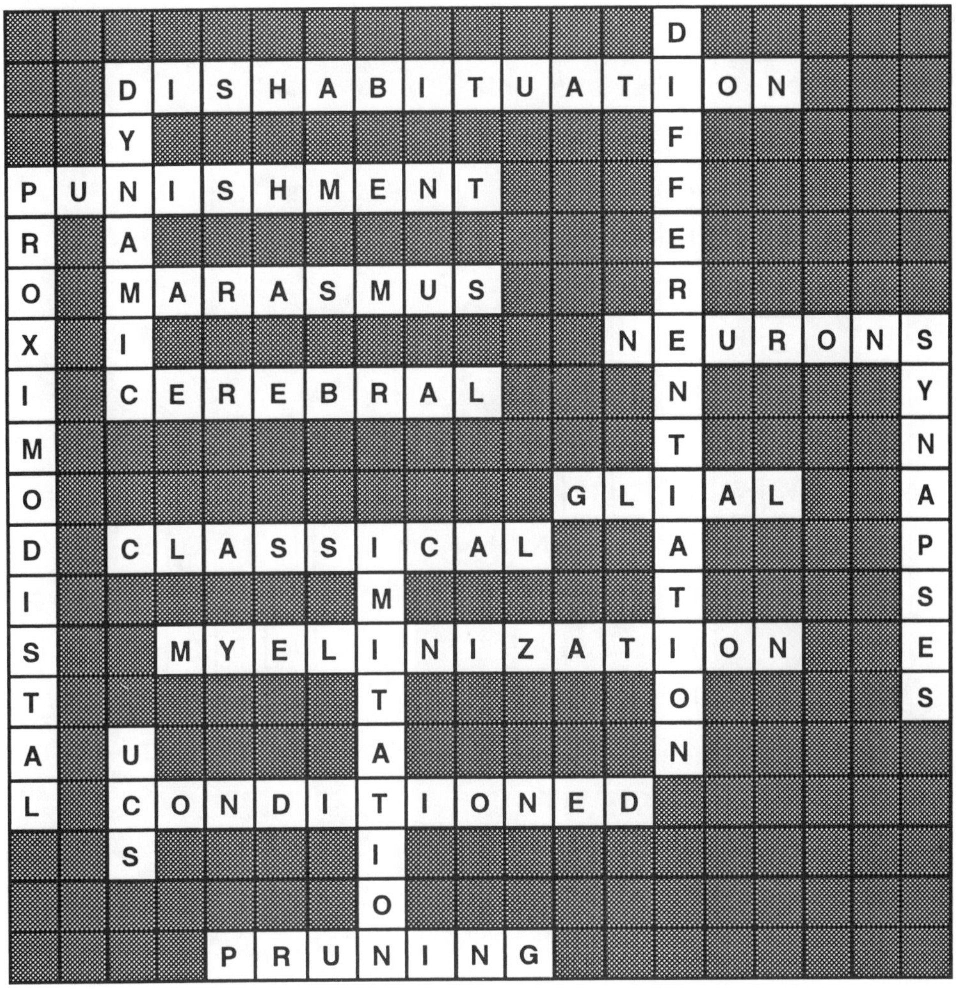
DISHABITUATION
PUNISHMENT
MARASMUS
NEURONS
CEREBRAL
GLIAL
CLASSICAL
MYELINIZATION
CONDITIONED
PRUNING
DIFFERENTIATION
DYNAMIC
PROXIMODISTAL
SYNAPSES
IMITATION
UCS

Puzzle 4.2

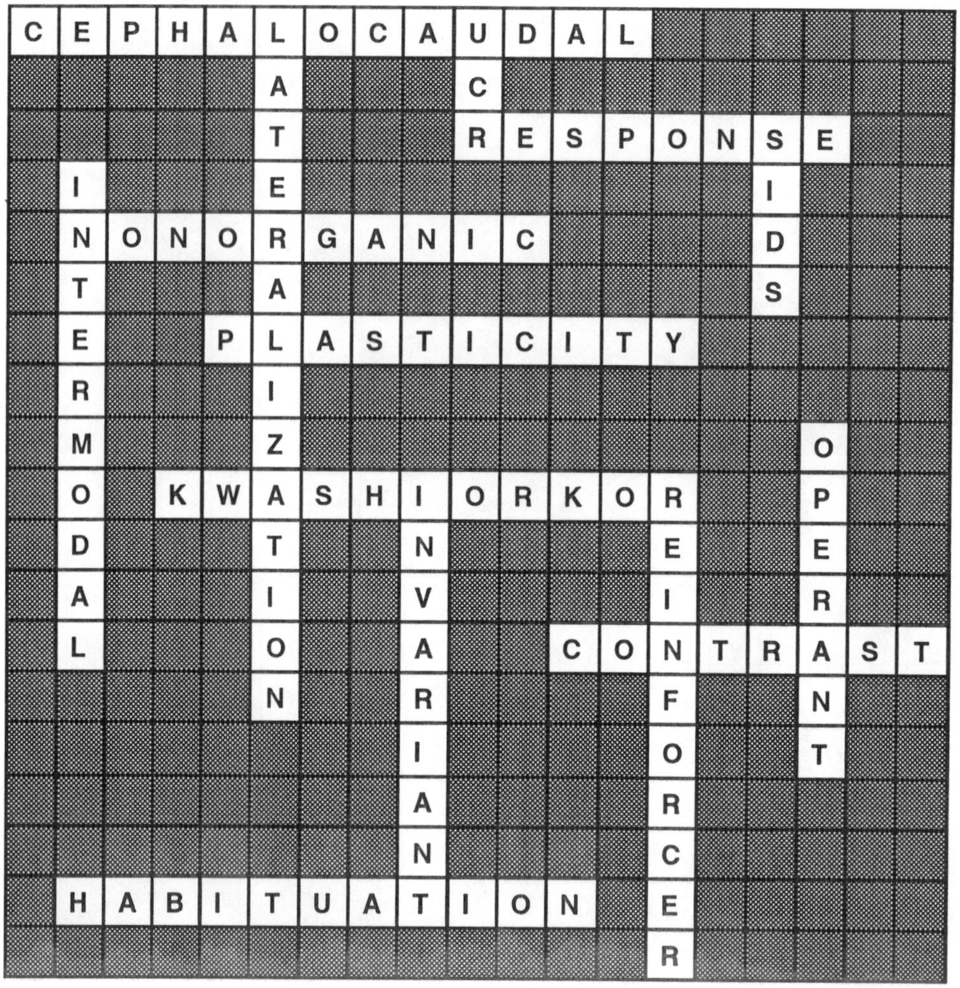
CEPHALOCAUDAL
ACRE
RESPONSE
SIDS
NONORGANIC
INTERMODAL
LATERALIZATION
PLASTICITY
KWASHIORKOR
INVARIANT
REINFORCER
OPERANT
CONTRAST
HABITUATION

Puzzle 5.1

A	C	C	O	M	M	O	D	A	T	I	O	N								
S															D		M			
S	E	N	S	O	R	I	M	O	T	O	R				E		E			R
I				R											F		N			E
M		R	E	G	I	S	T	E	R			O	B	J	E	C	T			C
I				A		C									R		A			O
L				N		H							C		R		L	O	N	G
A				I		E	X	P	R	E	S	S	I	V	E					N
T				Z		M							R		D					I
I		A		A		E							C							T
O		D		T									U							I
N		A		I	Q			C	A	U	S	A	L	I	T	Y				O
		P		O									A							N
I	N	T	E	N	T	I	O	N	A	L			R							
		A																		
A	U	T	O	B	I	O	G	R	A	P	H	I	C	A	L					
		I									O									
	W	O	R	K	I	N	G				M									
		N						M	A	K	E		B	E	L	I	E	V	E	
															A					
															D					

Puzzle 5.2

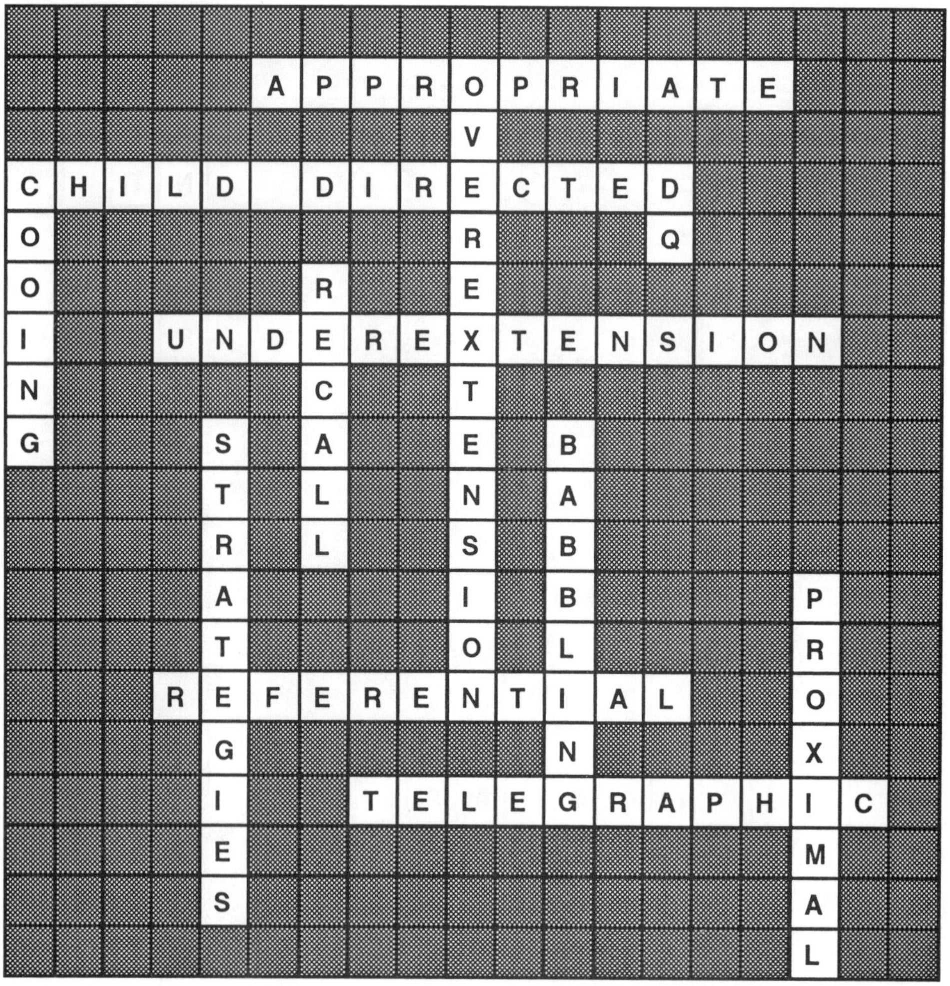
APPROPRIATE
CHILD DIRECTED
UNDEREXTENSION
REFERENTIAL
TELEGRAPHIC
COOING
OVEREXTENSION
DQ
RECALL
STRATEGIES
BABBLING
PROXIMAL

Puzzle 6.1

									B											
C				S	E	P	A	R	A	T	I	O	N		B	A	S	I	C	
O		E		Y					S							T			O	
M		M		N				R	E	S	I	S	T	A	N	T			N	
P		P		C						O						A			S	
L		A		H						C						C			C	
I	N	T	E	R	N	A	L			I		E				H			I	
A		H		O		V				A	U	T	O	N	O	M	Y		O	
N		Y		N		O		S		B		H				E			U	
C				Y		I		O		L		O				N			S	
E		T				D		C		E		L				T		D		
	S	E	L	F		A		I				O						I		
		M				N		A				G	O	O	D	N	E	S	S	
A		P				T		L				I						O		
N		E	A	S	Y					M		C		S	E	C	U	R	E	
X		R		T		S	M	I	L	E		A		H		O		I		
I		A		R								L		Y		N		E		T
E		M		A				S	L	O	W					T		N		R
T		E		N				E								R		T		U
Y		N		G				L								O		E		S
		T		E		D	I	F	F	I	C	U	L	T		L		D		T

Puzzle 7.1

						S	C	R	I	P	T	S		R	E	C	A	S	T	S
			P				H													O
O	R	D	I	N	A	L	I	T	Y											C
V			T				L						M	E	M	O	R	Y		I
E			U				D		P				E							O
R			I						E				T							D
R			T				C	A	R	D	I	N	A	L	I	T	Y			R
E			A						C				C			R				A
G			R						E				O			A				M
U			Y			A			P		E		G			N				A
L						N			T		X		N			S				T
A	C	A	D	E	M	I	C		I		P		I			F				I
R						M			O		A		T			O				C
I						I			N		N		I		O	R	T			
Z						S					S		O			M				
A		G	R	O	W	T	H				I		N			A				
T						I					O					T				
I						C			C	E	N	T	R	A	T	I	O	N		
O											S					O				
N																N				
							P	R	A	G	M	A	T	I	C	S				

Puzzle 7.2

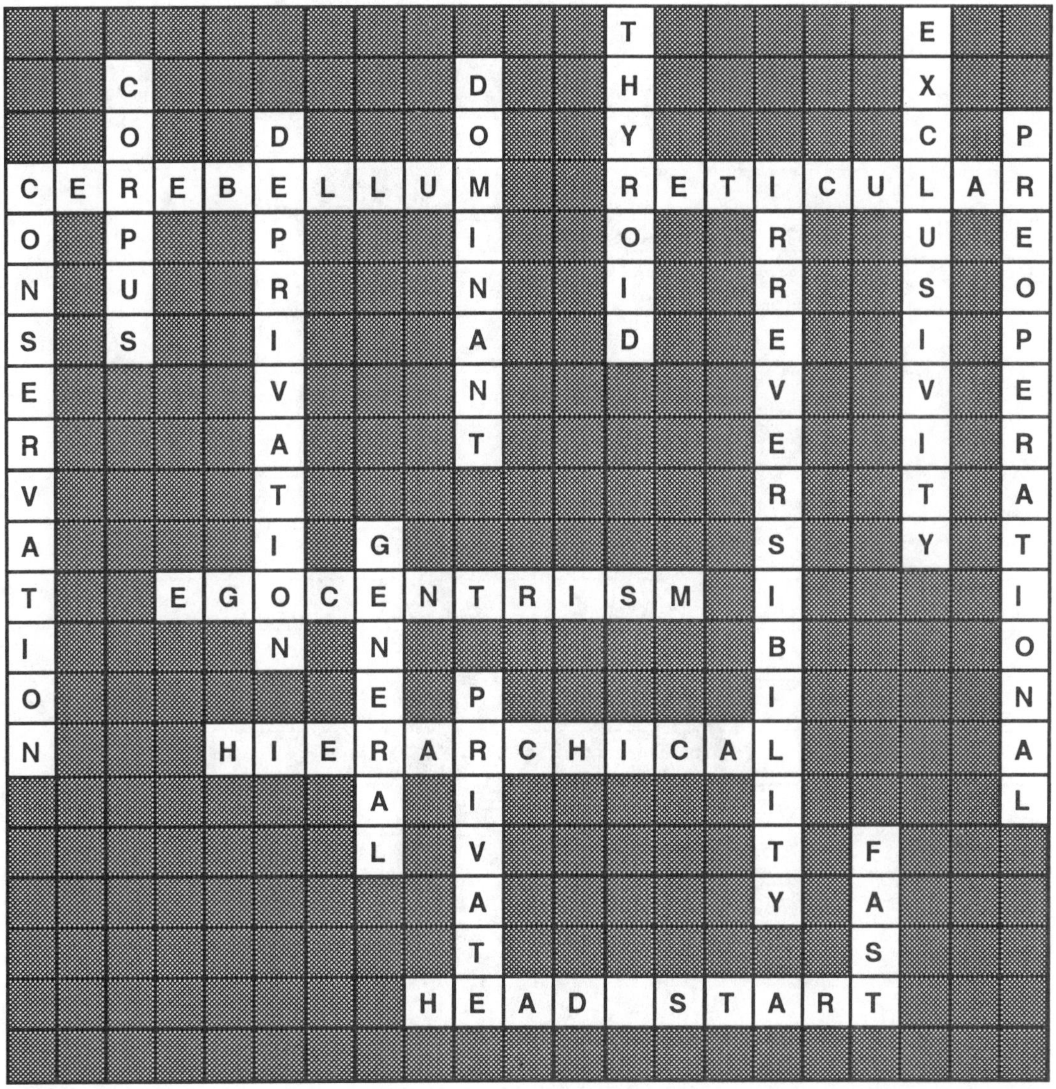

CEREBELLUM
RETICULAR
EGOCENTRISM
HIERARCHICAL
HEAD START
CONSERVATION
CORPUS
DEPRIVATION
DOMINANT
THYROID
IRREVERSIBILITY
EXCLUSIVITY
PREOPERATIONAL
GENERAL
PRIVATE
FAST

Puzzle 8.1

E	S	T	E	E	M		I	D	E	N	T	I	F	I	C	A	T	I	O	N
	Y																Y		V	
	M		A	U	T	H	O	R	I	T	A	T	I	V	E		P		E	
	P		U			O			N								I		R	
	A		T			S			I	N	D	U	C	T	I	O	N		T	
	T		H			T			T								G			
	H		O			I			I										P	
	Y		R			L			A	S	S	O	C	I	A	T	I	V	E	
			I			E			T				O		L		N		R	
			T					T	I	M	E		O	U	T		S		M	
A			A		S				V				P		R		T		I	
N			R		C	O	N	C	E	P	T		E		U		R		S	
D			I		H			O		A			R		I		U		S	
R			A		E			N		R			A		S		M		I	
O			N		M			S		A			T		T		E		V	
G					A			T		L			I		I		N		E	
Y								A		L			V		C		T			
N	O	N						N		E			E				A			
Y								C		L							L			
	I	D	E	N	T	I	T	Y												
										R	E	L	A	T	I	O	N	A	L	

Puzzle 9.1

						G						S						B		
C	O	G	N	I	T	I	V	E			R	E	H	E	A	R	S	A	L	
				N		F						R						S		
O		D		H		T		O	B	E	S	I	T	Y				I		T
P		E		I		E						A						C		R
E		C		B		D			M	U	L	T	I	P	L	E				A
R		A		I								I					M	A	I	N
A		L		T			C					O	P	E	N			D		S
T	R	A	D	I	T	I	O	N	A	L		N						H		I
I		G		O			N						E		T			D		T
O		E		N			V						L		R		E			I
N						D	E	C	E	N	T	R	A	T	I	O	N			V
A							R						B		A		U			E
L		D	I	V	E	R	G	E	N	T			O		R		R			
							E						R		C		E			
			L	E	A	R	N	I	N	G			A		H		S	E	L	F
	W						T						T		I		I			U
	H												I		C		S			L
	O		O	R	G	A	N	I	Z	A	T	I	O	N						L
	L												N			M	I	L	D	
R	E	V	E	R	S	I	B	I	L	I	T	Y								

Puzzle 10.1

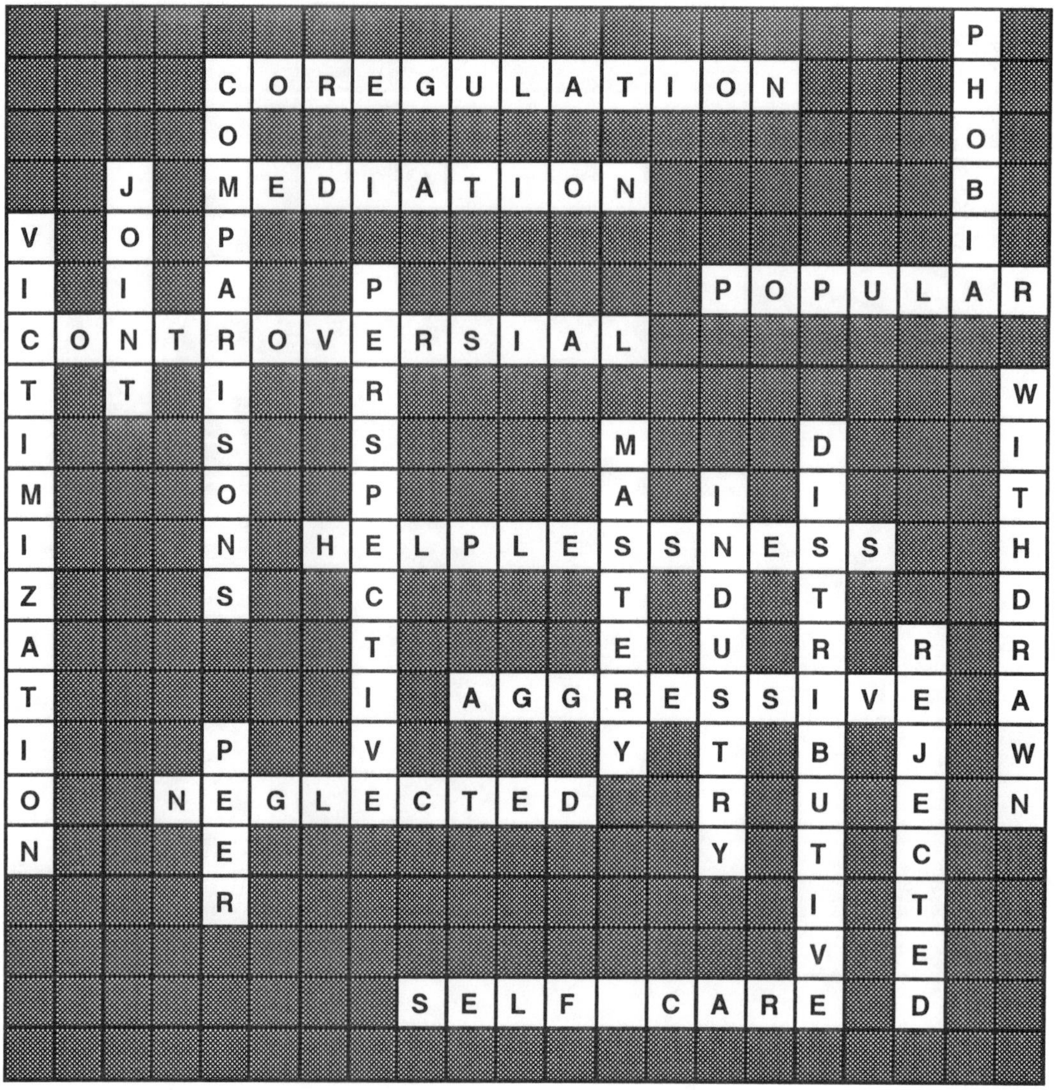

PHOBIA
COREGULATION
COMPARISONS
JOINT
MEDIATION
VICTIMIZATION
PERSPECTIVE
POPULAR
CONTROVERSIAL
WITHDRAWN
MASTERY
DISTRIBUTIVE
INDUSTRY
HELPLESSNESS
REJECTED
AGGRESSIVE
PEER
NEGLECTED
SELF CARE

Puzzle 11.1

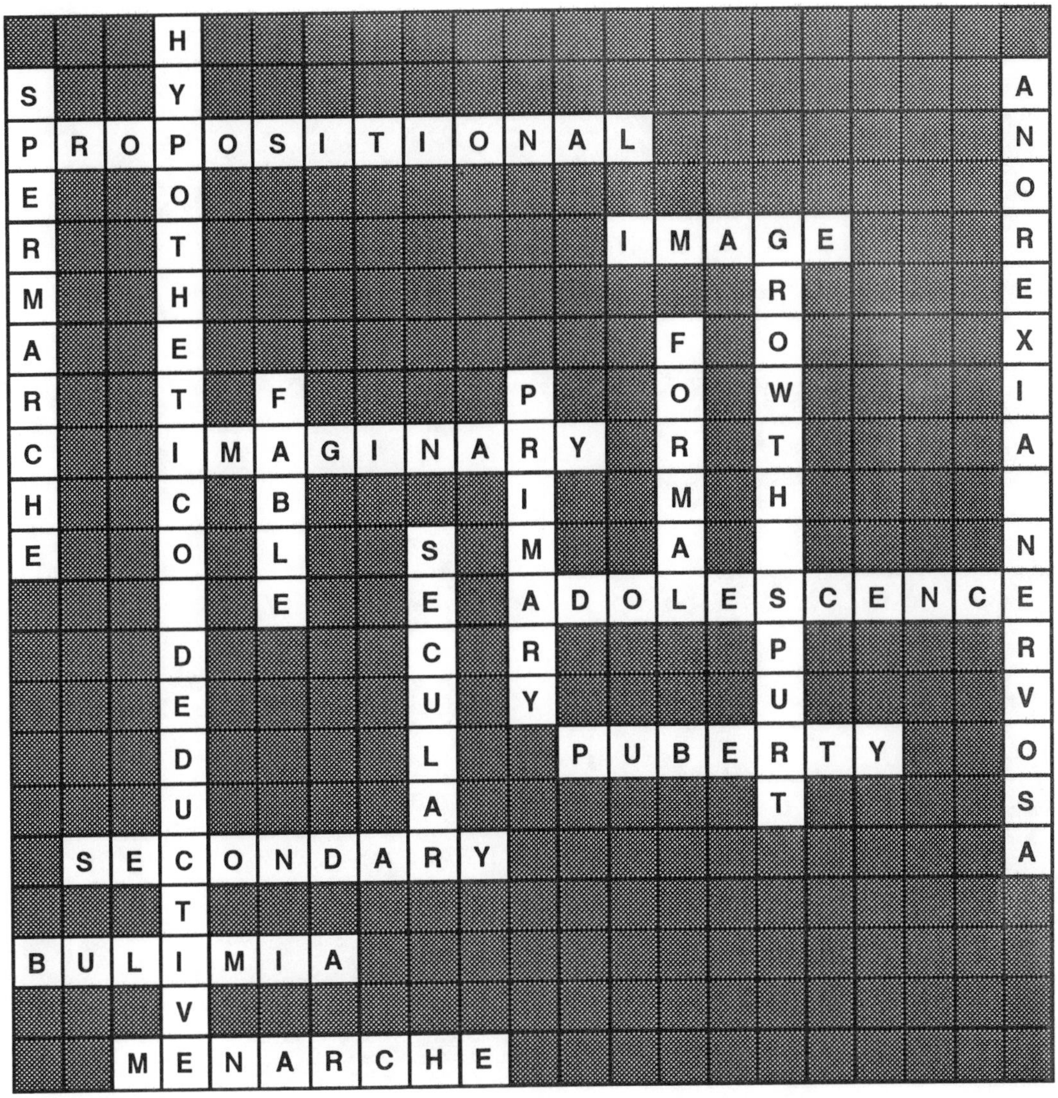
PROPOSITIONAL
IMAGE
IMAGINARY
ADOLESCENCE
PUBERTY
SECONDARY
BULIMIA
MENARCHE
SPERMARCHE
HYPOTHETICO
DEDUCTIVE
FABLE
PRIMARY
SECULAR
FORMAL
GROWTH
SPURT
ANOREXIA
NERVOSA

Puzzle 12.1

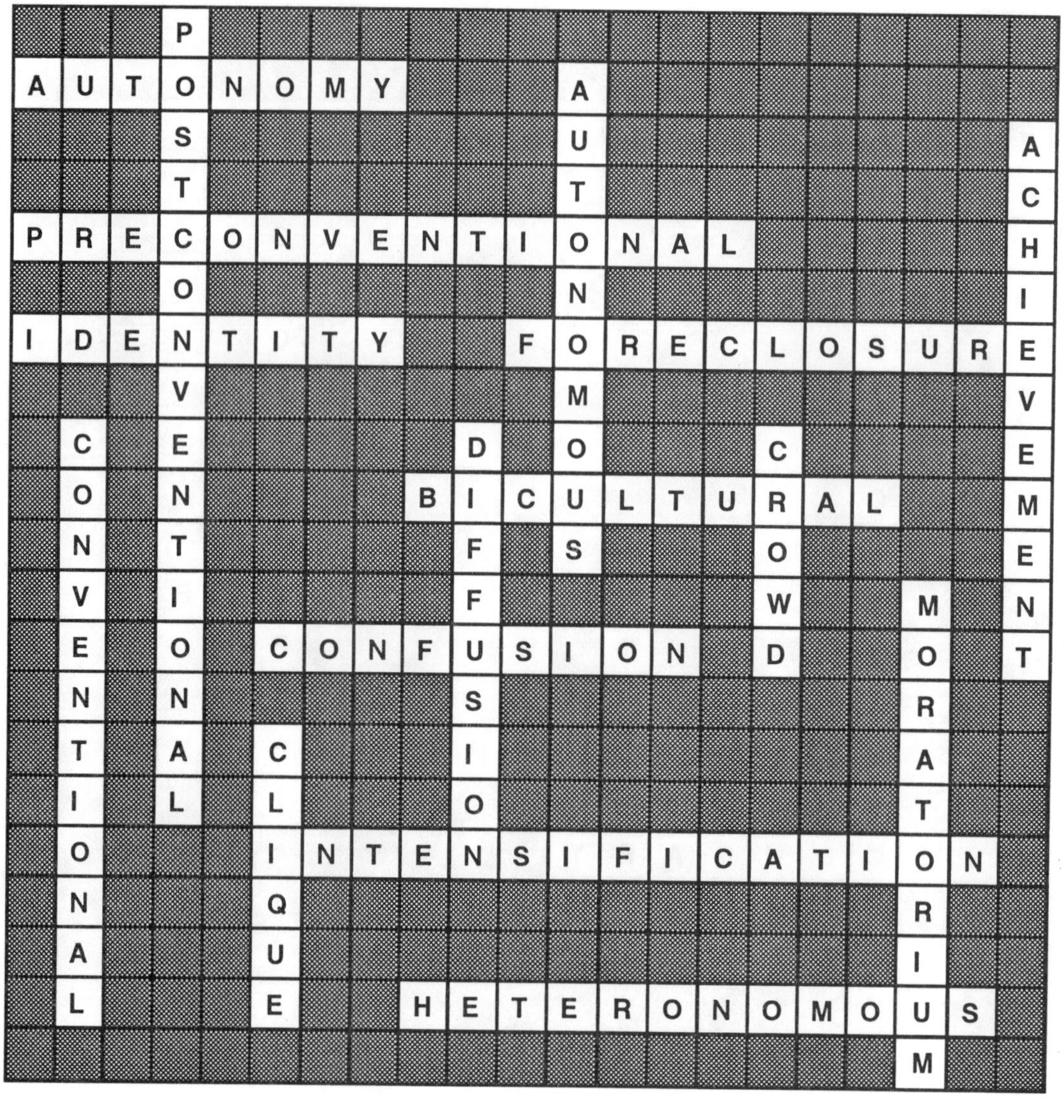
AUTONOMY
PRECONVENTIONAL
IDENTITY
FORECLOSURE
BICULTURAL
CONFUSION
INTENSIFICATION
HETERONOMOUS
POSTCONVENTIONAL
AUTONOMOUS
ACHIEVEMENT
CONVENTIONAL
DIFFUSION
CROWD
MORATORIUM
CLIQUE

Puzzle 13.1

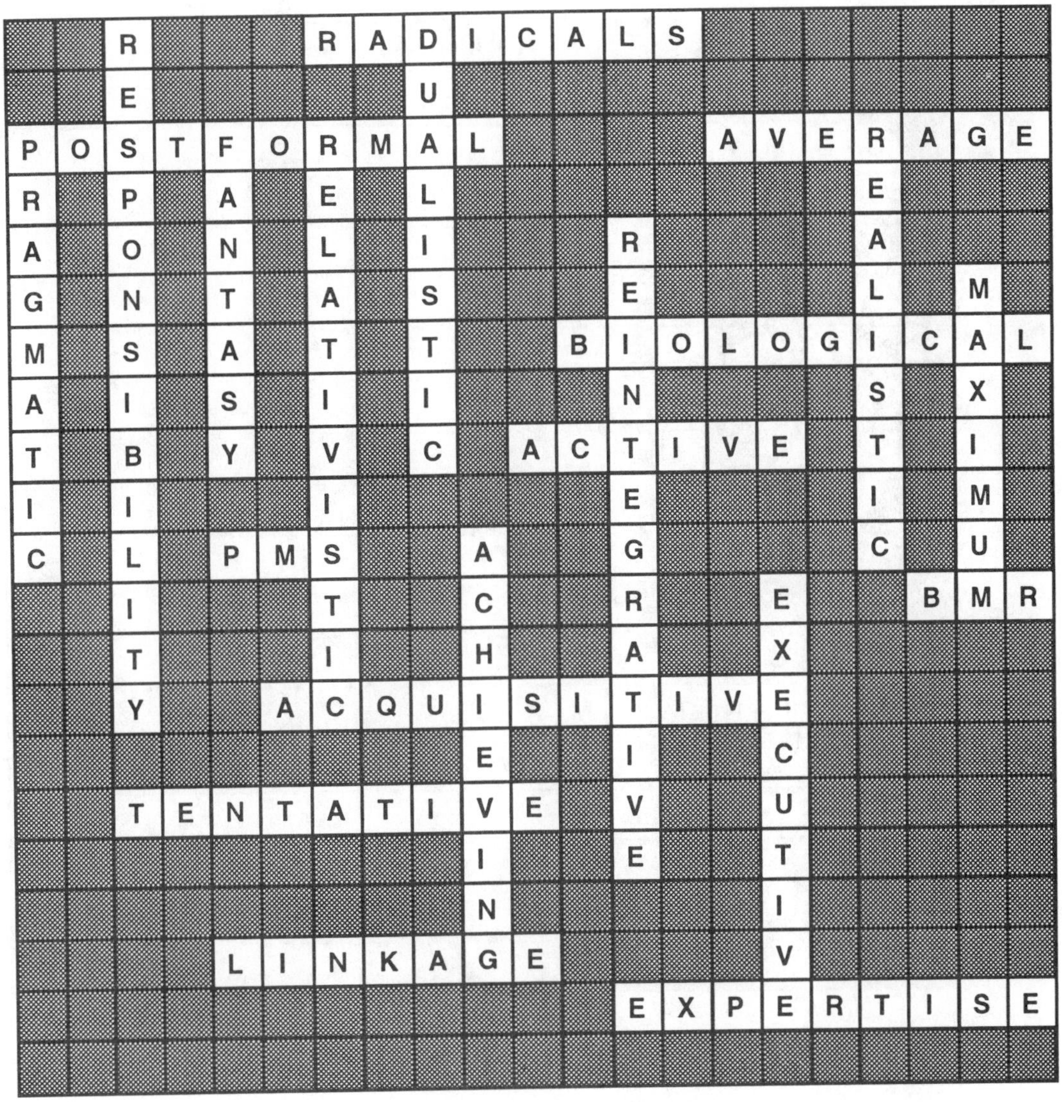
RADICALS
POSTFORMAL
AVERAGE
BIOLOGICAL
ACTIVE
PMS
BMR
ACQUISITIVE
TENTATIVE
LINKAGE
EXPERTISE
RESPONSIBILITY
DUALISTIC
PRAGMATIC
FANTASY
RELATIVISTIC
REINTEGRATIVE
REALISTIC
MAXIMUM
ACHIEVING
EXECUTIV

Puzzle 14.1

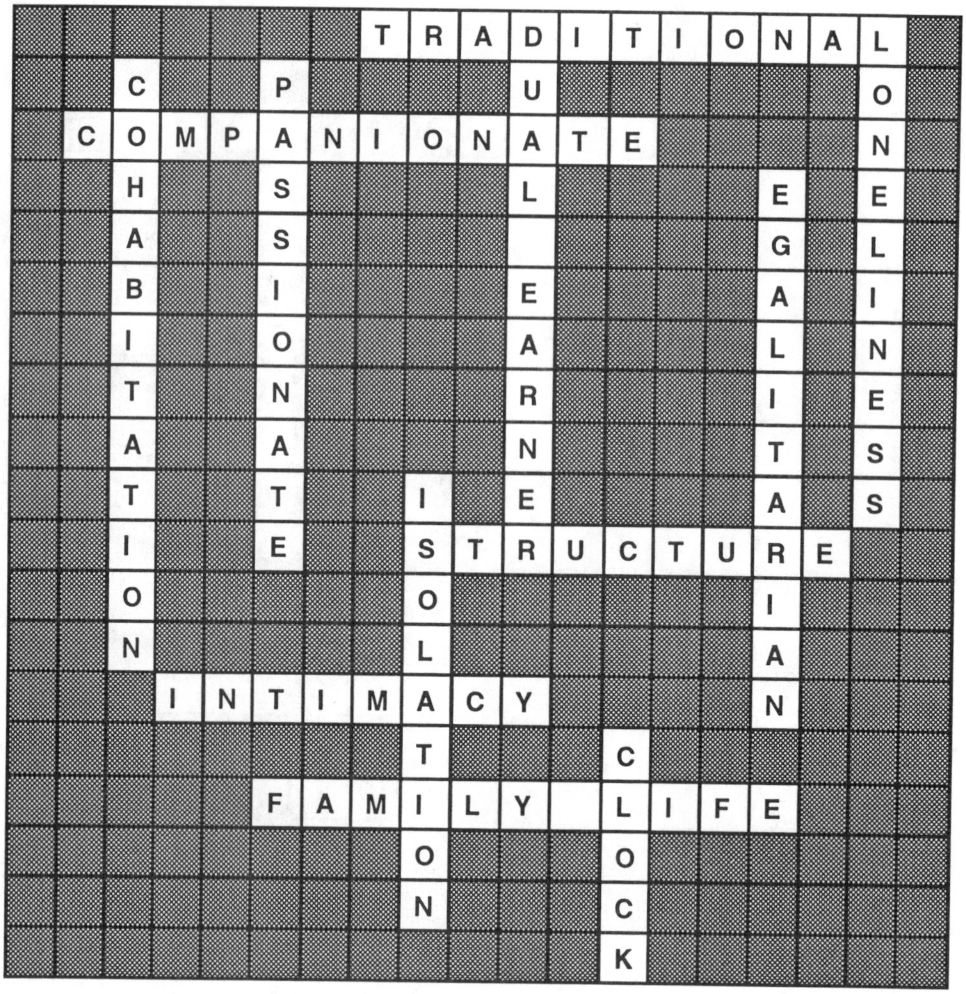
TRADITIONAL
COMPANIONATE
COHABITATION
PASSIONATE
DUAL EARNER
EGALITARIAN
LONELINESS
ISOLATION
STRUCTURE
INTIMACY
FAMILY LIFE
CLOCK

Puzzle 15.1

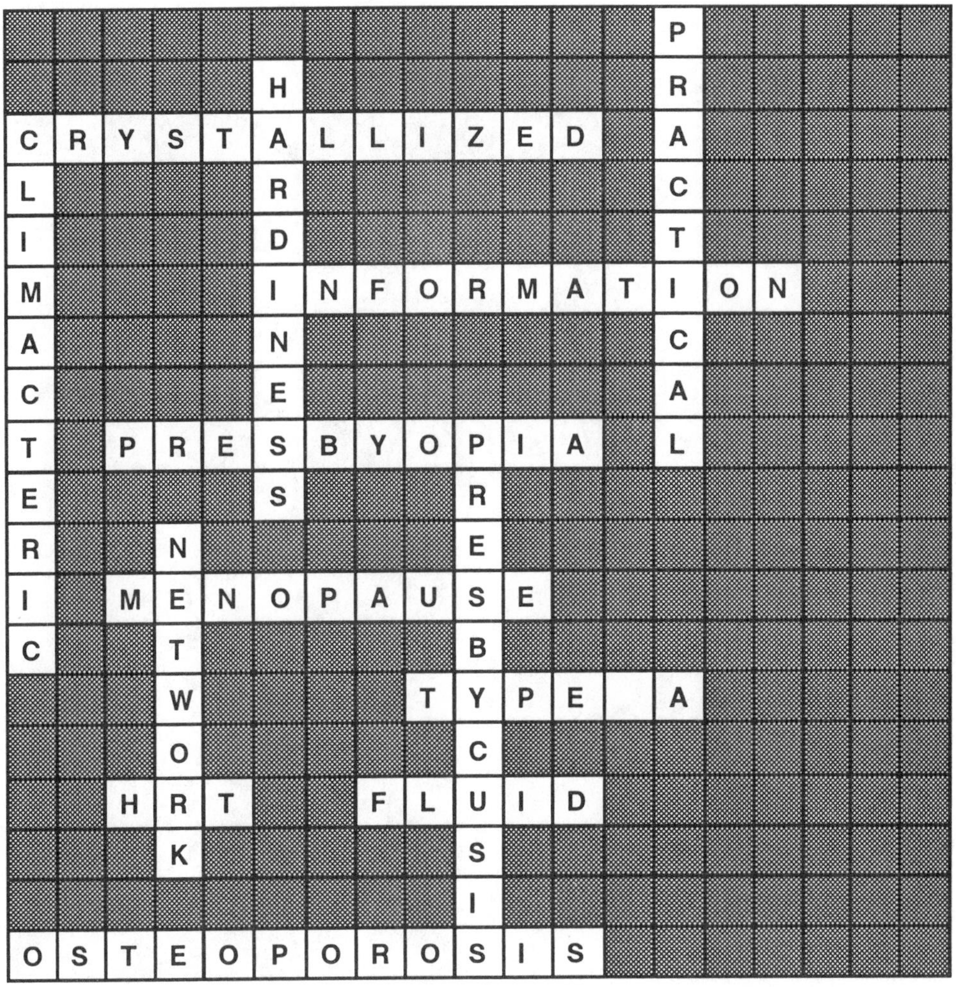
CRYSTALLIZED
INFORMATION
PRESBYOPIA
MENOPAUSE
TYPE A
HRT
FLUID
OSTEOPOROSIS
PRACTICAL
HARDINESS
CLIMACTERIC
PRESBYCUSIS
NETWORK

Puzzle 16.1

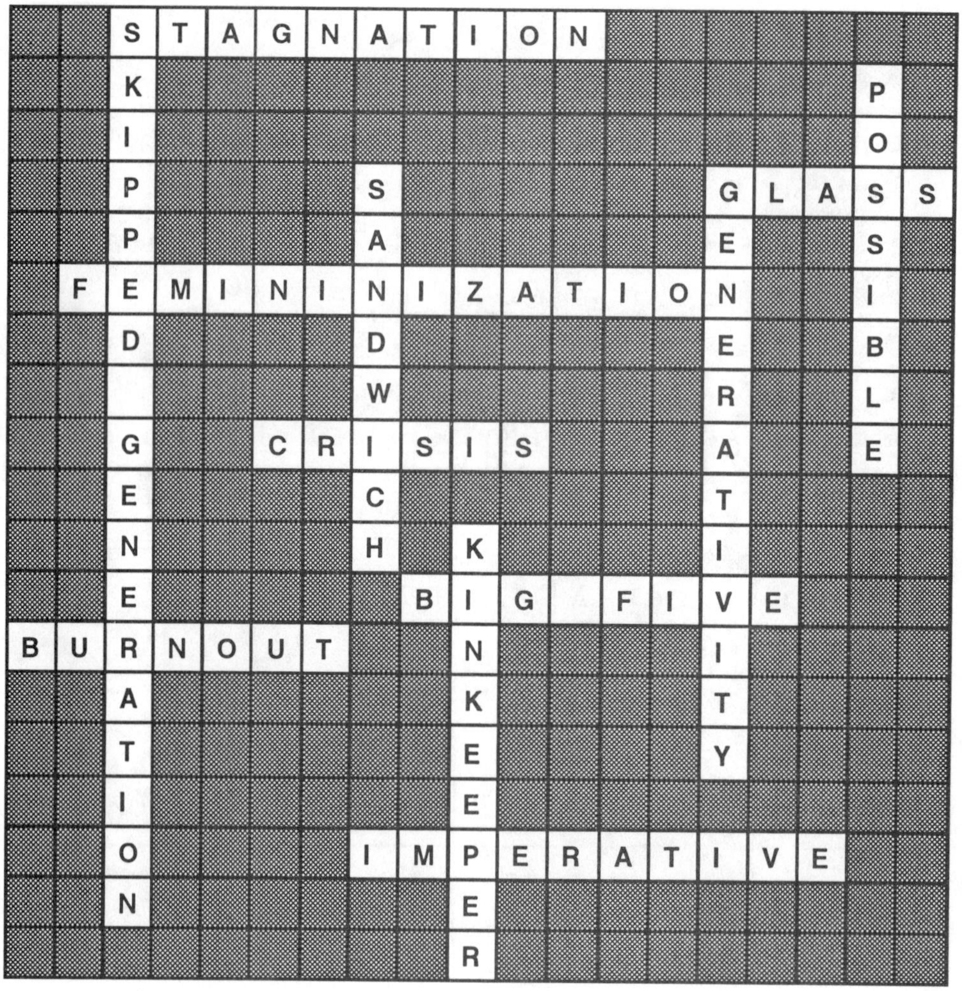
STAGNATION
SKIPPED
POSSIBLE
SANDWICH
GLASS
FEMINIZATION
GENERATIVITY
CRISIS
GENERATION
KINKEEPER
BIG FIVE
BURNOUT
IMPERATIVE

Puzzle 17.1

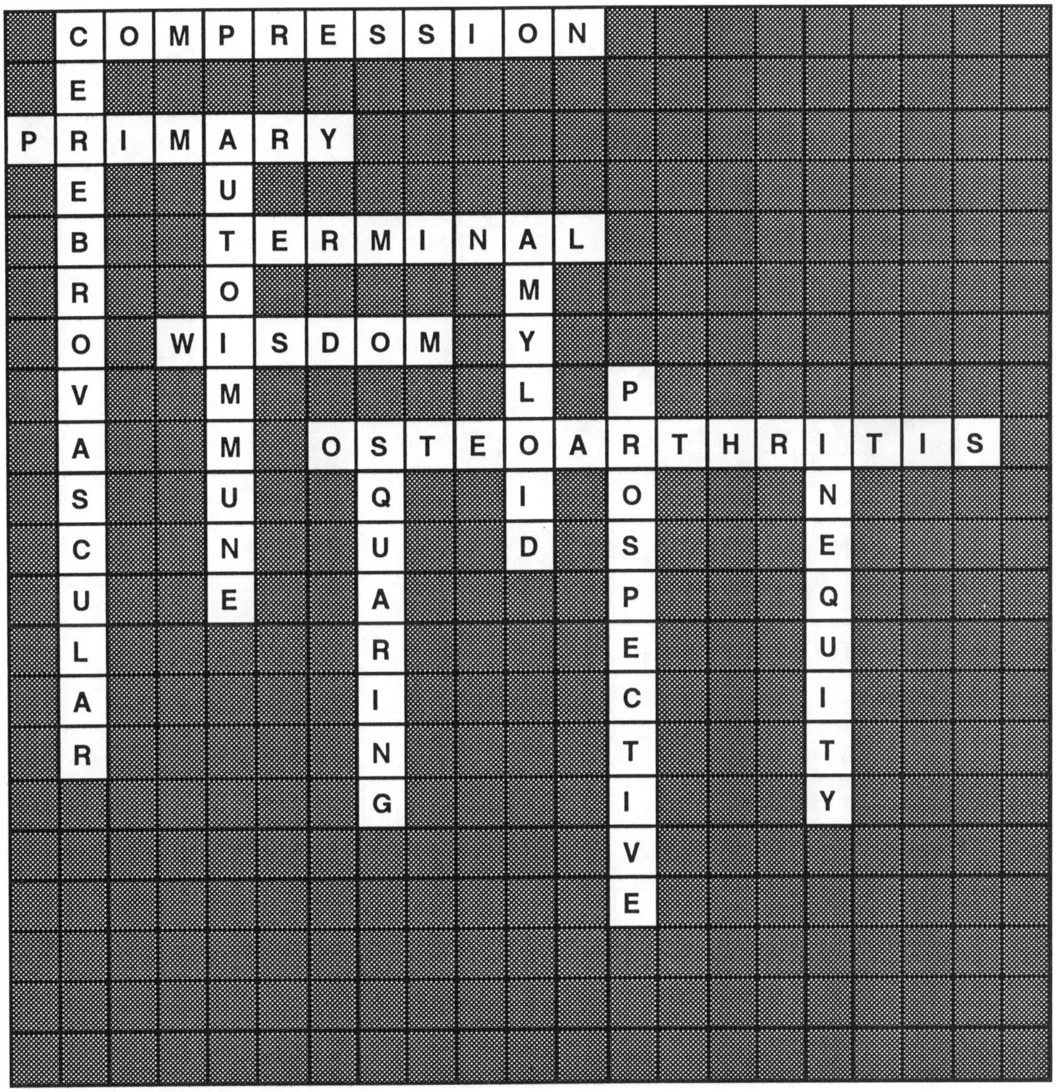

COMPRESSION
CEREBROVASCULAR
PRIMARY
AUTOIMMUNE
TERMINAL
AMYLOID
WISDOM
PROSPECTIVE
OSTEOARTHRITIS
SQUARING
INEQUITY

Puzzle 17.2

N	E	U	R	O	F	I	B	R	I	L	L	A	R	Y				
			H						M									
		R	E	M	O	T	E		P						F			
			U						L						U			
			M						I				C		N			
			A	P	N	E	A		C	A	T	A	R	A	C	T	S	
			T						I				O		T		E	
			O	L	D				T				S		I		C	
			I										S		O		O	
			D	E	M	E	N	T	I	A			O		N		N	
													V		A		D	
													E		L		A	
					A	L	Z	H	E	I	M	E	R	S			R	
											A						Y	
											C							
									Y	O	U	N	G					
											L							
				O	P	T	I	M	I	Z	A	T	I	O	N			
											R							

Puzzle 18.1

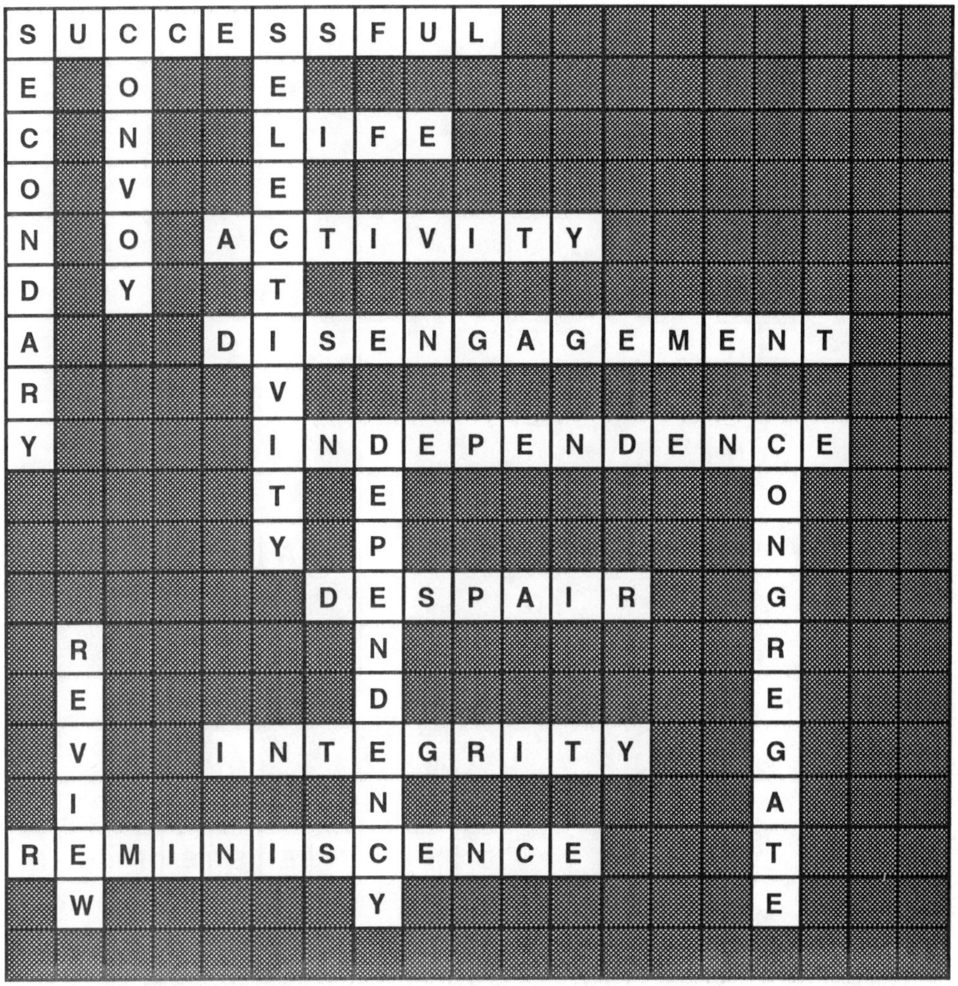
SUCCESSFUL
SECONDARY
CONVOY
SELECTIVITY
LIFE
ACTIVITY
DISENGAGEMENT
INDEPENDENCE
DEPENDENCY
CONGREGATE
DESPAIR
REVIEW
INTEGRITY
REMINISCENCE

Puzzle 19.1

				A						A				A	G	O	N	A	L	
		A	C	C	E	P	T	A	N	C	E				R					
		P		T		E				C					I					
		P		I		R			H	O	S	P	I	C	E					
		R		V		M				M					F					
		O		E		A				M										
		P				N				O			A							
		R				E			A	D	V	A	N	C	E					
		I				N				A			G					E		
B	R	A	I	N		C				T			E					U		
A		T				E		U	N	I	V	E	R	S	A	L	I	T	Y	
R		E								O								H		
G						D			A	N	X	I	E	T	Y			A		
A	T	T	O	R	N	E	Y									P		N		
I						N										A		A		
N						I										S		S		
I						A					D	E	P	R	E	S	S	I	O	N
N						L										I		A		
G																V				
								B	E	R	E	A	V	E	M	E	N	T		

Puzzle 19.2

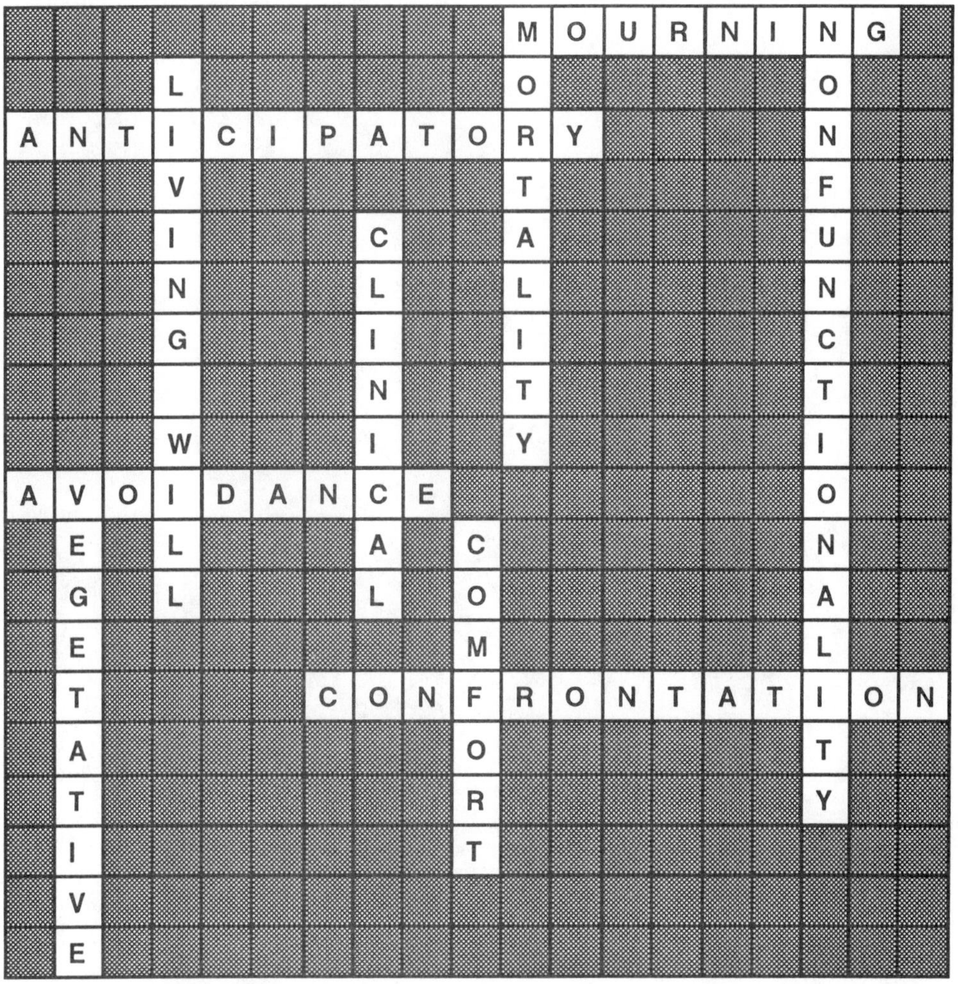
MOURNING
LIVING WILL
ANTICIPATORY
MORTALITY
NONFUNCTIONALITY
CLINICAL
AVOIDANCE
VEGETATIVE
COMFORT
CONFRONTATION

ANSWERS TO SELF-TESTS

Chapter 1

1. d	2. b	3. b	4. d	5. d
6. c	7. c	8. b	9. a	10. c
11. a	12. d	13. c	14. c	15. d
16. b	17. b	18. b	19. a	20. d
21. a	22. a	23. b	24. c	25. a

Chapter 2

1. b	2. c	3. c	4. a	5. d
6. c	7. c	8. b	9. d	10. c
11. d	12. a	13. b	14. b	15. d
16. a	17. c	18. d	19. a	20. c
21. c	22. a	23. d	24. a	25. c

Chapter 3

1. a	2. c	3. b	4. d	5. a
6. d	7. a	8. a	9. c	10. d
11. c	12. b	13. c	14. c	15. b
16. b	17. a	18. d	19. d	20. d
21. b	22. c	23. b	24. b	25. c

Chapter 4

1. d	2. b	3. a	4. c	5. b
6. b	7. d	8. a	9. a	10. c
11. b	12. d	13. d	14. d	15. c
16. c	17. a	18. b	19. c	20. b
21. d	22. a	23. b	24. a	25. a

Chapter 5

1. b	2. c	3. a	4. b	5. b
6. c	7. d	8. a	9. c	10. a
11. a	12. b	13. c	14. c	15. c
16. a	17. a	18. b	19. b	20. c
21. c	22. a	23. b	24. c	25. a

Chapter 6

1. d	2. c	3. a	4. d	5. b
6. c	7. b	8. b	9. a	10. b
11. a	12. c	13. c	14. b	15. c
16. b	17. a	18. d	19. c	20. d
21. d	22. a	23. c	24. b	25. d

Chapter 7

1. c	2. a	3. a	4. c	5. a
6. d	7. b	8. b	9. a	10. c
11. b	12. b	13. b	14. c	15. a
16. b	17. c	18. b	19. d	20. d
21. a	22. b	23. c	24. b	25. d

Chapter 8

1. b	2. d	3. b	4. d	5. c
6. a	7. c	8. c	9. b	10. c
11. a	12. c	13. c	14. b	15. a
16. d	17. b	18. d	19. c	20. b
21. a	22. b	23. a	24. c	25. b

Chapter 9

1. a	2. b	3. a	4. c	5. b
6. b	7. c	8. d	9. c	10. a
11. c	12. c	13. a	14. d	15. b
16. d	17. b	18. c	19. b	20. c
21. c	22. c	23. c	24. a	25. b

Chapter 10

1. d	2. c	3. c	4. c	5. d
6. c	7. d	8. b	9. a	10. b
11. c	12. d	13. b	14. c	15. b
16. b	17. d	18. b	19. c	20. a
21. c	22. d	23. b	24. a	25. d

Chapter 11

1. c	2. c	3. c	4. d	5. c
6. d	7. d	8. c	9. c	10. a
11. b	12. d	13. a	14. d	15. a
16. b	17. a	18. c	19. b	20. a
21. b	22. c	23. a	24. a	25. a

Chapter 12

1. d	2. c	3. c	4. a	5. c
6. b	7. a	8. c	9. b	10. b
11. d	12. b	13. a	14. a	15. d
16. c	17. a	18. b	19. c	20. a
21. b	22. b	23. c	24. b	25. a

Chapter 13

1. a	2. d	3. c	4. a	5. b
6. c	7. d	8. d	9. c	10. c
11. b	12. a	13. d	14. c	15. c
16. b	17. c	18. c	19. c	20. a
21. a	22. b	23. d	24. c	25. d

Chapter 14

1. d	2. b	3. d	4. a	5. d
6. c	7. b	8. b	9. c	10. b
11. d	12. d	13. a	14. c	15. d
16. b	17. b	18. c	19. c	20. d
21. d	22. c	23. a	24. b	25. d

Chapter 15

1. c	2. c	3. d	4. d	5. d
6. b	7. b	8. d	9. a	10. d
11. a	12. b	13. c	14. d	15. d
16. b	17. a	18. a	19. d	20. a
21. c	22. d	23. a	24. a	25. d

Chapter 16

1. d	2. b	3. d	4. a	5. d
6. a	7. d	8. d	9. d	10. a
11. d	12. c	13. a	14. d	15. d
16. a	17. c	18. c	19. b	20. a
21. b	22. c	23. d	24. c	25. d

Chapter 17

1. c	2. d	3. c	4. b	5. d
6. b	7. d	8. b	9. d	10. c
11. b	12. a	13. d	14. a	15. c
16. d	17. c	18. a	19. b	20. a
21. c	22. d	23. d	24. c	25. b

Chapter 18

1. d	2. d	3. a	4. b	5. c
6. d	7. a	8. d	9. b	10. c
11. b	12. a	13. b	14. c	15. d
16. c	17. d	18. b	19. d	20. c
21. b	22. b	23. a	24. b	25. d

Chapter 19

1. a	2. d	3. d	4. a	5. d
6. a	7. c	8. d	9. b	10. b
11. c	12. d	13. d	14. c	15. c
16. b	17. d	18. d	19. b	20. d
21. c	22. b	23. b	24. a	25. d